The Florida Perennial Garden

Easy Plants That Thrive in Heat, Sand & Humidity

Jermaine Jefferson

GrowFitFL LLC

This book offers information solely for educational purposes. Educational. While the author has tried to ensure accuracy, gardening practices, environmental conditions, and results may vary by location, climate, and individual circumstances. The author and publisher make no guarantees regarding outcomes and assume no responsibility for errors, omissions, or damages resulting from the use of this information.

Published by
GrowFitFL LLC
www.growfitfl.com

ISBN
Paperback ISBN: 979-8-9947732-0-8

Printed in the United States of America

Contents

Part 1: The Florida Rules of Engagement

READ THIS FIRST: WHY THIS BOOK EXISTS

L et me start by saying this. I love perennial plants.

If you are following my YouTube channel, @GrowFitFL, you know perennials are a top video topic for our backyard garden community.

Honestly, I've never understood why anyone living in Florida would choose anything but perennials. No offense! We have year-round growing conditions. Heat is present. We have rain. Why would you restart your garden from seed every year when you could plant something once and harvest from it for the next decade?

When you have a house full of people to feed, this just always made more sense to me.

And I know I'm not alone in this.

How do I know? My YouTube channel has hundreds of videos on Florida gardening. The most viewed, most commented-on, most shared videos? They're all about perennials. People are hungry for this information. If you're reading this book, you probably feel the same way.

The problem I ran into early on, when I was setting up my backyard food forest, was information. Or rather, the lack of it.

Nobody I could find broke down why perennials are so powerful in Florida. Nobody explained why they're our golden ticket to year-round abundance with a fraction of the work. I did not know how many perennials actually thrive here. I didn't know which ones to avoid, or how to keep them alive past the first summer.

Perennials are the best low-maintenance investment in your garden.

Think about it. Low maintenance. Drought tolerant. Heat tolerant. Truly, most farmers can set these crops and forget them.

Annuals are the opposite. Daily watering, pest protection, fertilizing, replanting every season. More work, every single year.

Learning about perennials completely blew my mind!

I learned through trial and error. A lot of errors.

This book is everything I wish I had had ten years ago. It's a decade of hands-on experience, dead plants, breakthroughs, and hard-won knowledge distilled into a system that actually works in Florida's unique conditions.

But I have to warn you first. Florida is just different.

You already know this. You have likely killed some plants by following advice that worked perfectly in North Carolina or California, or wherever the gardening book originated or a non-Florida YouTube channel you have been following.

Florida has its own rules of engagement. Rules you have to follow before, during, and after planting to make sure those precious perennials don't become annuals by your own hand.

That's what this book is for.

By the time you finish reading and applying what's inside, you're going to have a garden that produces food year-round with less work than most people put into mowing their lawn. You're going to understand why things work here instead of just copying someone else's plant list and hoping for the best.

And when it works, and it will, I want to hear about it.

Tag me @GrowFitFL on YouTube, Facebook, TikTok, wherever. I can't wait to see your gardens thriving, your harvests piling up, and your confidence growing.

I'm known for no fluff on my YouTube channel and podcast, so let's bring that same energy here.

No filler. No theory that doesn't work in the real world. Just the information you need to succeed.

Let's get started.

Chapter 1 Forget Everything You Know

WHY TRADITIONAL GARDENING ADVICE FAILS IN FLORIDA. THE FLIPPED SEASONS, THE BRUTAL SUMMERS, AND WHY SUCCESS HERE REQUIRES UNLEARNING BEFORE LEARNING.

Having taken the advice, you planted in the spring. You changed the soil. You provided consistent watering. And everything died anyway.

It was not your fault.

Florida does not follow the rules you learned in gardening books, YouTube videos, or advice from relatives up north. Florida flipped the growing calendar. The soil is not actually soil. Summer is not the growing season. It is survival mode. And the plants that thrive everywhere else fail here spectacularly.

If you have ever felt like you are a terrible gardener because nothing works here, you are not alone. You just need different rules.

This chapter is about unlearning what you think you know so you can start fresh with what actually works.

The Calendar Is Backwards

In most of the United States, spring is planting season. Gardens come alive in April and May. Summer is when everything grows. Fall is harvest time. Winter is dormancy.

Florida laughs at this entire concept.

Spring, March through May, is not the beginning of the growing season. It is the countdown to hell. Temperatures climb into the 90s. Humidity spikes. Afternoon thunderstorms roll in like clockwork. You are racing against time to get plants established before summer stress tests everything you just put in the ground.

Summer, June through September, is when plants struggle to survive. Heat, humidity, and daily rain create the perfect environment for fungus, root rot, and pest explosions. Plants that loved full sun everywhere else wilt here by noon. Watering does not help because the problem is not drought. It is too much of everything at once.

Fall, October and November, is when Florida actually wakes up. Temperatures drop into the 70s and 80s. Humidity eases. The relentless afternoon storm tapers off. This is the proper planting season. This is when roots establish without fighting heat stress.

Winter, December through February, is not dormancy. It is prime growing season for most vegetables and many perennials. North Florida gets some frost, but Central and South Florida stay warm enough that plants keep growing straight through what the rest of the country calls the off-season.

Plant tomatoes in March, like the seed packets say, and they will bolt before June. Wonder why your herbs die in May? It is because they are trying to reproduce before the summer heat kills them. Treat winter like nothing grows, and you'll miss the best part of the Florida gardening year.

Someone flipped the calendar here. Accept that or keep losing plants on schedule.

Heat plus humidity equals failure

Everyone knows Florida is hot. But the heat here is not the same as the heat in Arizona or Southern California.

A 95-degree day in Phoenix is dry. Plants cool themselves through transpiration. Water evaporates. Air moves.

A 95-degree day in Orlando is 95 degrees with 80 percent humidity and an afternoon thunderstorm that dumps two inches of rain in thirty minutes. The air does not move. Transpiration barely works. The soil stays wet for hours. Fungus blooms overnight.

Therefore, full-sun plants from other regions fail here. They cannot handle heat plus humidity and standing water all at the same time.

Tomatoes that thrive in California develop fungal diseases here in days. Lavender rots at the roots. Roses drown in black spot, powdery mildew, and aphid infestations.

Assume heat tolerance in one climate equals heat tolerance in Florida, and you will replace plants every season.

Your soil does not exist

What is in your yard right now is not soil. It is sugar sand or crushed limestone that drains like a sieve and holds nutrients about as well as a colander holds water.

Before we built our food forest, and - just sugar sand.

Real soil has structure. It has organic matter, microbes, fungi, earthworms, and clay particles that hold on to water and nutrients long enough for plants to use them.

Florida sand has none of that.

Add compost to sand, and it breaks down in three to six months. Not years. Months. The heat and moisture here speed up decomposition so fast that organic matter disappears before it builds proper soil structure.

If you are on limestone-based soil in South Florida, the problem is distinct but equally frustrating. The pH is too high. Micronutrients get locked out. Hardpan layers form beneath the surface and block root growth. Add sulfur, and the limestone fights back to alkaline within months.

You cannot treat Florida sand like Midwestern loam. Build soil from nothing, and that takes years, not weekends.

Ignore this, and you will spend hundreds of dollars on amendments that wash away with the next thunderstorm. We will cover how to build real soil in Chapter 3.

Watering Kills More Plants Than Drought

Here is where most people go wrong.

They observe that a plant is wilting. People assume that the plant requires water. They provide water for them. The plant dies anyway.

In Florida, more water is not the answer. Drainage is.

Plants do not die here from lack of water. They die from too much water sitting around the roots, creating anaerobic conditions that invite root rot and fungal infections.

You can water a plant every single day and still lose it if the water does not drain away fast enough.

It rains constantly in summer, but drainage matters more than irrigation. You can have a yard that floods every afternoon and still need to improve soil drainage, or your plants drown in saturated sand.

Consistent watering schedules do not work here either. Watering every morning at 7 AM sounds disciplined, but it ignores what actually happened overnight. Did it rain two inches at 3 AM? Then you do not need to water. Did the storm miss your yard entirely? Then maybe you do.

Florida rewards observation, not routine.

Learn to read plants. Is the soil still wet two inches down? Do not water. Are the leaves drooping at 10 AM before the heat peaks? That is heat stress, not drought. Is the plant wilting at 5 PM after a dry week? Now it needs water.

Follow a timer instead of watching the plant, and you will kill everything that does not thrive on neglect.

Unlearn before you can learn

Everything you thought you knew about gardening does not apply here. The calendar is reversed. The heat is different. The soil does not exist. Watering is counterintuitive. And most people writing gardening advice online assumed climates with nothing in common with Florida.

But once you understand Florida's actual rules, gardening here gets easier, not harder.

Perennials that match this climate do not just survive. They thrive with less work than annuals require anywhere else. You do not need to fight Florida. You just need to stop trying to make it behave like North Carolina.

The plants in this book work with Florida's climate instead of against it. They handle heat and humidity. In sand, they grow. They shrug off summer storms. They do not need you to recreate a different climate in your backyard.

But before we get to plant selection, you need to know which Florida you are in.

Because the rules change depending on whether you are in Pensacola, Orlando, or Miami.

Let us figure that out next.

Chapter 2 Find Your Florida

NORTH, CENTRAL, AND SOUTH FLORIDA EXPLAINED CLEARLY.

N orth, Central, or South. The rules change depending on which one you're in. Get this wrong and you'll waste hundreds of dollars replacing plants that would never survive your winters.

Florida isn't one climate. It's three completely different growing environments that share a state border.

Plant selection that works in Jacksonville fails in Fort Lauderdale. Timing that works in Tampa confuses people in Tallahassee. Get this wrong and you'll spend years replacing expensive tropicals that would never survive your winters.

I'm in west-central Florida. My backyard sits in the gambling zone. Some years we get frost. Some years we don't. I've lost plants to freeze I didn't see coming, and I've pushed tropicals that thrived for three years before dying in one cold snap. My wife keeps telling me to stop planting mangoes, keep them in pots, and she's been right twice now.

31 degrees = cold damage. The tree will recover.

This chapter is about knowing which Florida you're in before you plant anything. Not memorizing USDA zone maps. Understanding what your specific region allows you to grow and what it'll kill no matter how hard you try.

By the end, you'll know which plants you can grow confidently, which ones are a waste of money in your region, and why your neighbor's mango thrives while yours freezes.

The Three Floridas

Florida divides into three distinct regions based on winter temperatures, frost frequency, and how tropical you can actually go with your plant selection.

Your region sets the boundaries for what's possible. Then microclimates fracture those boundaries at the yard level.

North Florida: You get frost

This is Pensacola to Jacksonville, everything north of Ocala.

USDA Zones: 8b to 9a

You get frost most winters. Sometimes hard freezes.

Your last frost happens mid-February to early March. Your first frost arrives mid-November to early December. That gives you a frost-free window, but it also means you can't treat your yard like the tropics.

Citrus is risky without protection. Papayas will fruit for three years, then die overnight in a single freeze. Moringa planted in the ground is a gamble you'll lose eventually.

But here's the advantage: winter dormancy resets pests and diseases. Plants that need a chill period to fruit or flower perform well here. You have access to a wider palette than Central or South Florida because you can grow borderline temperate plants that would struggle further south.

The challenge? Borderline tropical plants are a slow-motion disaster. They thrive for years, lull you into confidence, then die in one cold snap.

Backup plans are necessary. You need container strategies. You need to accept that some plants will always be seasonal here, not permanent.

Plant tropicals in the ground and you'll have to replant them every few years. Count on it.

Central Florida: The Gambling Zone

This is Ocala to Sarasota, I-4 corridor, and everything in between.

USDA Zones: 9b to 10a

Some years you get frost. Some years you don't. This is the gambling zone.

You get occasional light frost from late December to early February, but it's inconsistent. One winter might hit 28 degrees and kill everything tender. The next winter might stay above 40 the entire season.

You never know which winter you're getting until it's too late.

Your neighbor three streets over has a mango that's been producing for a decade. You plant the same variety. Yours freezes two years in. Cold air sinks into your low-lying lot while theirs stays warm near the lake.

Same county. Different Florida.

I've seen this happen in my neighborhood. One guy's papaya is eight feet tall and fruiting. Mine died to the ground in January when we hit 29 degrees for four hours a few years back. His yard is near a retention pond. Mine sits in a low spot where cold air settles. We're less than half a mile apart.

The advantage: you can grow almost anything. This is the widest plant palette in the state. Tropicals thrive most years. Subtropical plants are bulletproof. Even some temperate perennials tolerate the mild winters.

The challenge: every planting decision is a calculated risk.

You cannot assume safety. You can't assume disaster. Certainty is not what you are betting on; it's probability.

Treat Central Florida like South Florida, and a single freeze costs you hundreds of dollars in dead plants. Treat it like North Florida, and you miss out on years of growth you could've had.

The tension never goes away.

Most people guess wrong. They either overprotect and limit themselves or under-protect and lose everything in one terrible winter.

The key is knowing which plants are worth the gamble and which ones aren't.

South Florida: You're Tropical

This is Fort Myers to Key West, everything south of Lake Okeechobee.

USDA Zones: 10b to 11

Frost is a once-in-a-decade event that makes the news when it happens. You're tropical. Forget temperate advice entirely.

You have no frost dates. Winter means daytime highs in the 70s and nighttime lows in the 60s. Summer means relentless heat and humidity with no break. Plants don't go dormant. They grow twelve months a year.

Bananas produce year-round. Moringa grows faster than you can prune it. Papayas fruit nonstop. Cassava, katuk, longevity spinach, Okinawa spinach—these plants don't just survive here, they dominate.

Meanwhile, blueberries that need winter chill refuse to fruit no matter what you do. Apples won't set. Peaches are impossible.

The advantage: year-round growing for tropical perennials. You can harvest something every single month without replanting. Production never stops.

The challenge: heat and humidity are relentless. Nothing resets. Pests don't die back in winter. Fungi never take a break.

Soil organic matter breaks down faster than anywhere else in the state. Mulch disappears in months. Compost vanishes. You're feeding the sand on a continuous loop.

Managing constant growth becomes your routine. Pruning, harvesting, controlling size. There's no off-season.

You either keep up or get buried.

Fight the climate and you exhaust yourself trying to grow temperate crops that would never work. Lean into the tropics and you harvest more food with less effort than gardeners anywhere else in the country.

Which Florida are you? The 30-Second Quiz.

Answer these three questions.

1. Do you see frost most winters?

Yes: North Florida

Sometimes: Central Florida

Almost never: South Florida

2. Can you grow citrus without protection?

Risky even with protection: North Florida

Yes, but freeze scares happen: Central Florida

Year-round without worry: South Florida

3. What's your biggest climate concern?

Occasional cold snaps: North Florida

Unpredictable freezes: Central Florida

Relentless heat with no winter break: South Florida

Here's the trick question most people get wrong.

4. If your yard has never frozen, are you safe planting tropicals in the ground?

Most people say yes.

The correct answer is no if you're in Central or North Florida.

One freeze in ten years is enough to kill a papaya, a moringa, or a banana to the ground. Just because it hasn't happened yet doesn't mean it won't happen.

If you're in South Florida, the answer is yes. If you're in Central or North Florida, the answer is containers or replanting.

What Your Region Means for Plant Selection

If you're in North Florida

Plant tropicals in containers so you can move them during freezes. A single night below 32 degrees kills a moringa in the ground. In a pot, you roll it into the garage and save it.

Take advantage of dormancy. It's your pest control reset button. Many perennial pests and diseases can't survive a hard freeze. Use that.

Fall planting works differently here than in Central or South Florida. Your fall is shorter and your winter is colder. Pay attention to frost dates and don't push tender plants into November.

Focus on zone 8 to 9 perennials as your backbone. Use tropicals as seasonal accents or container plants, not permanent landscape features.

Ignore frost dates, and you'll replace expensive plants every spring. Accept them, and you build a garden that lasts.

If you're in Central Florida,

Diversify. Plant some tropicals. Plant some subtropical. Keep frost cloth on hand for the years it freezes.

Frost cover over a small mango tree.

This is the most forgiving region for experimentation. You can trial plants that would die immediately in North Florida or struggle in South Florida's relentless heat.

Your neighbors' yards will look wildly different based on microclimates. Someone three streets over never sees frost because they're near a lake. You get hit every other year because cold air sinks into your low-lying lot.

Watch your specific yard. Not the county average.

One terrible winter will teach you more about your property than five mild ones. Pay attention when it happens.

In my backyard, I've learned which beds freeze and which ones stay warmer. The bed near the house never drops below 35. The bed near the back fence hits 28 almost every cold snap. Same yard. Ten-degree difference.

If you're in South Florida

Stop trying to grow plants that need winter chill.

Apples, peaches, and blueberries that require cold hours won't fruit here no matter what you do. Let them go.

Lean into true tropicals. Moringa, katuk, cassava, longevity spinach, Okinawa spinach, papayas, bananas. These plants produce year-round with minimal input.

Your challenge isn't triggering growth—it's managing constant growth. Plants don't go dormant here. They grow twelve months a year. That means pruning, harvesting, and managing size become part of your routine.

Soil building is constant because organic matter breaks down faster here than anywhere else in the state. Mulch disappears in months. Compost vanishes.

You're feeding the sand on a continuous loop.

Fight the climate and you exhaust yourself. Work with it and you will harvest more food with less effort than gardeners anywhere else in the country.

Microclimates Fracture Every Region

Your official USDA zone is a starting point, not a guarantee.

Because water moderates temperature swings, a yard near a lake stays warmer in winter. A yard on high ground with good airflow might see frost when the valley a mile away stays warm. A yard surrounded by concrete and buildings creates a heat island that pushes you half a zone warmer.

Pay attention to your specific property, not just the regional average.

Walk your yard in winter and note where cold air settles. That low spot near the fence will frost before the raised bed near the house. Plant accordingly.

Watch where water pools after a storm. That's your drainage problem area. Avoid putting anything there that hates wet feet.

Notice which areas get blasted by afternoon sun and which get shade from structures or trees. Full sun in Florida isn't the same as full sun in Colorado.

Adjust your plant selection.

Your yard has its own rules. Learn them.

I figured this out in year two after moving into our home. I noticed our driveway bed never freezes because the concrete radiates heat all night. My wife planted ginger there. I planted ginger in the back bed. She survived every cold snap. Mine died twice before I moved it.

You Can't Change Your Florida, So Work With It

You're in North, Central, or South Florida. You can't change that.

You can fight it by trying to grow plants meant for a different region, replacing them every year, and wondering why nothing works.

Or you can accept your Florida climate , choose plants that match it, and let your garden thrive.

The next section organizes plants by their success requirements. Heat tolerance, shade tolerance, salt tolerance (yes, that's a thing) and which Florida they actually work in.

But none of that matters if you plant them wrong.

Because what you think is soil in your yard isn't soil at all. And until you understand what you're actually planting into, every plant you put in the ground is fighting a losing battle from day one.

Part 2: The Plants That Actually Work

Chapter 3 Your Soil Doesn't Exist Yet (Here's How to Build It)

WHAT YOU'RE REALLY PLANTING INTO. SUGAR SAND, LIMESTONE, AND HOW PH BEHAVES HERE. REFRAMING SOIL BUILDING AS A PROJECT THAT PAYS OFF OVER TIME.

If you moved here from anywhere else, you looked at your yard and thought; *I need to amend this soil.*

Wrong.

What's in your yard isn't soil. It's sugar sand or crushed limestone that drains like a sieve and holds nutrients about as well as a colander holds water.

You can't amend what doesn't exist. Build it from scratch.

This isn't your fault. This is physics.

My wife, Toni, is a realtor, and I've accompanied her to plenty of open houses over the years.

One thing I've learned: if you're a gardener looking for a new home, you can truly appreciate another gardener who's selling and leaving you with good soil.

Real soil, the kind with structure, organic matter, microbes, fungi, earthworms, and clay particles that hold water and nutrients, doesn't happen overnight in Florida.

It takes years of intentional work.

When you walk into a property with established fruit trees, amended beds, and a functioning food forest, you're looking at tens of thousands of dollars in value and decades of labor that most buyers will never see.

I told Toni: if we ever move, I'm only selling to a gardener. Someone who will appreciate what we've built. Someone who will use what we've done to Grow Food NOT a Lawn.

Real soil has structure.

Organic matter. Microbes, fungi, earthworms, and clay particles that hold onto water and nutrients long enough for plants to use them. Florida sand has none of that. It never will on its own. No amount of waiting or wishing changes this.

Sandy soil in my Central Florida backyard garden.

You're not planting into soil. You're planting into a substrate that was never designed to grow anything.

Until you accept that, you'll keep losing plants and blaming yourself.

Here's what I wish someone had told me ten years ago when I started my food forest in west-central Florida. You are standing on soil that doesn't need fixing. It's the absence of soil.

And once you understand that, everything changes.

You stop trying to fix something that won't change and start creating something new from scratch. It takes time. It takes patience. But it works.

By the end of this chapter, you'll know what Florida sand actually is, why traditional soil advice fails here, and the minimum viable strategy for building real soil that improves every year instead of washing away with the next storm.

What Florida "Soil" Actually Is

Most of Florida sits on one of two materials. Sugar, sand, or limestone.

Both drain fast. Neither holds nutrients. The advice that works in real soil doesn't apply to either.

When I first moved to Florida, I kept reading about soil amendments and thinking, *okay, I just need to add more compost.* More peat moss. More organic matter.

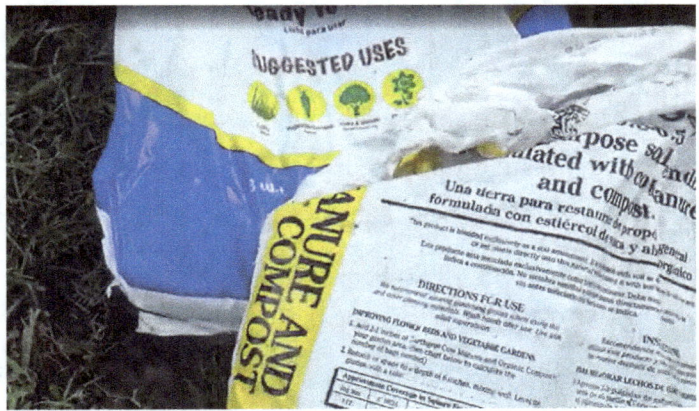

Hundreds of dollars in peat moss and cow manure later, I finally learned the right way to build Florida soil.

I added truckloads.

They vanished.

For a long time, I assumed I was failing. I adjusted techniques, added more material, spent more money, and watched the same thing happen again and again. Eventually, effort was not the problem. I was not dealing with the soil. I was dealing with geology.

Much of Florida gardening begins on sugar sand. Nearly pure quartz. It feels like beach sand because it essentially is. Water moves straight through it. Nutrients follow. There is no structure to slow anything down, no capacity to hold what you add.

Sugar sand does not improve simply because time passes. It does not reorganize itself. It does not quietly become soil while you wait. Sand is a stable material. Left alone, it remains exactly what it is.

Many people assume nature will fix it if given enough patience. That belief makes sense elsewhere. In Florida, it usually leads to frustration. Compost disappears. Amendments vanish. The ground feels unchanged because it is unchanged.

This is why so many gardeners add organic matter season after season and still feel sand when they dig. Not because they did something wrong. Because sand does not transform without a system that keeps material in place long enough to matter.

Central and North Florida sit almost entirely on this foundation. If you dig and it looks like a shoreline, that is your reality. The work begins with accepting that fact, not resisting it.

Further south, the challenge shifts. There, the ground is limestone. Crushed coral and shells compressed into a hard, alkaline base. Water drains quickly, but chemistry becomes the dominant force.

High pH locks away micronutrients. Iron, manganese, and zinc can all be present yet inaccessible. Plants show distress even when fertilized correctly. Leaves yellow. Growth stalls. The issue is not absence. It is availability.

This leads many gardeners into an endless cycle of correction. Add sulfur. Add iron. Adjust. Watch improvement. Then watch it fade. Limestone does not yield easily. It's buffering power overwhelms small interventions.

In some places, solid layers form beneath the surface. Roots meet resistance. Water collects above it. Growth becomes constrained not by care, but by what lies below.

Clay exists in scattered pockets farther north, bringing the opposite problem. Water lingers. Oxygen disappears. Roots suffer. Clay has structure, but drainage becomes the battle instead of retention.

These realities explain why traditional gardening advice often falls short here.

Most guidance assumes soil already exists. Add compost. Add mulch. Build from there. In Florida, organic matter breaks down so quickly that it rarely stays long enough to create structure on its own.

Heat and moisture speed up decomposition. What might last years elsewhere can disappear in months. Compost feeds biology briefly, then exits the system. Without protection, without layering, nothing accumulates.

This creates the illusion of progress followed by disappointment. People keep adding material, wondering why nothing seems to change. They are replacing what vanishes, not building what remains.

PH behaves the same way. Lime lifts acidity until rain moves it out. Sulfur lowers alkalinity until limestone pushes it back. We can manage numbers, but rarely control them permanently.

Fertilizer moves differently, too. Sand offers no buffer. Nutrients either wash away or concentrate. Excess burns roots quickly. Deficiency returns just as fast. Slow release becomes a necessity, not a preference.

What changes everything is not speed, but method.

Real soil does not arrive suddenly. It appears gradually, almost quietly. Texture shifts. Water lingers a little longer. Life moves in from the surface downward.

The biggest mistake is mixing organic matter into sand. Tilling feels productive, but it speeds up loss. Oxygen and heat speed up decomposition. Structure never forms because decomposition breaks it down before it can organize.

Surface layering works differently. Organic matter rests on top. Mulch shields it. Biology moves it downward at its own pace. This mirrors how soil forms in undisturbed systems.

Mulch stops being decoration and becomes infrastructure. Wood chips, leaves, and pine straw all feed the process as they break down. Rocks do not. Rubber does not. Fabric blocks the very organisms needed to do the work.

We used fabric once. It looked clean. It felt controlled. Later, we removed it and rebuilt properly. That correction mattered more than most additions we ever made.

Feeding plants alone keeps them alive. Feeding the soil allows the system to improve. Biology unlocks nutrients, moderates moisture, and builds a structure that does not wash away.

Tests can provide a baseline, but plants speak continuously. Yellowing with green veins signals iron lockout. Stunted growth reveals deeper chemistry. Leaves respond faster than paperwork ever will.

Success does not announce itself loudly. It shows up in fewer inputs, steadier growth, soil that smells alive, and ground that holds together when disturbed.

Florida gardening works when you stop asking the land to behave like somewhere else and start building a system that respects exactly what is already there.

Mulch breaking down with visible signs of biology working.

Growth often looks unimpressive at first. Plants settle in, but they do not surge. Above ground, progress seems modest. Below the surface, roots are expanding, adjusting, learning how to survive in unfamiliar conditions.

This phase is easy to misread. It feels like stagnation when it is actually preparation. Pushing harder here usually causes damage. Extra fertilizer does not speed up adaptation. It stresses roots that are still figuring out how to function in their new environment.

Nothing is wrong when growth appears restrained. That restraint is part of the process. The system is organizing itself before it expresses anything outwardly dramatic.

Most people give up here because it looks like nothing is happening. In reality, nearly everything is happening out of sight. Roots are claiming space. Microbial life is moving in. The foundation is being laid quietly.

Later, the soil itself changes. The surface no longer behaves like loose sand. It holds together slightly. It feels less gritty, more cohesive. Organic material is no longer disappearing immediately. It is mixing in, staying present, becoming part of the structure.

That shift is subtle, but it is decisive. Once the ground responds, the system has crossed from survival into stability.

Year 2 of organic matter mixing into
the sandy Florida soil.

Water lingers a little longer. Not enough to feel dramatic, just enough to notice. The ground does not dry out instantly. Moisture remains present instead of vanishing.

Perennials respond first. Their roots have reached far enough to do more than survive. Growth strengthens. Foliage thickens. The plants stop reacting to stress and start expressing potential.

The soil itself gives the clearest signal. Texture shifts. What once slipped through your fingers now holds together briefly. Watering becomes less frequent. Feeding becomes more effective.

Eventually, there is no question anymore. The top layer behaves like real soil. It smells right. It breaks apart the way soil should. Nutrients no longer disappear immediately. What you add stays long enough to be used.

Plants perform the way they do in established gardens elsewhere. Growth becomes steady. Production increases. The system stops resisting every input.

The space feels different. Beds that once struggled now support abundance. Care becomes lighter. Results become heavier.

I noticed the change slowly. My wife noticed it instantly. She walked through the beds one morning, pressed the ground with her hand, and said it out loud.

This does not feel like sand anymore.

And she was right.

My wife - Year 3 in our backyard garden.

She was right. It didn't.

Building soil isn't fast, but it's possible

You can't rush this. You can't buy your way out of it with expensive amendments.

Commit to the process and let time do the work.

But the perennials in the next section don't need perfect soil to start. They just need you to stop fighting the sand and start working with it.

These plants thrive in Florida's conditions as they are. These grow in sand. They tolerate heat and humidity. They don't require you to recreate a different climate in your backyard.

But even the toughest perennial will fail if you plant it wrong.

And in Florida, planting correctly means understanding drainage, spacing, and timing in ways that don't apply anywhere else.

Miss any of those and everything you just learned about soil building becomes irrelevant. The plant dies before the soil matters.

Let's talk about the mindset that makes all of this work.

The Florida Perennial Mindset: Think Like the Climate

FLORIDA DOESN'T RESPOND TO CONTROL. IT RESPONDS TO OBSERVATION. THIS IS THE SHIFT THAT MAKES EVERYTHING ELSE WORK.

D o you know why traditional gardening fails in Florida and which Florida zone you're in? And do you understand that what's in your yard isn't soil yet, but it can be.

Now we need to talk about the shift that makes everything else work.

Most people approach Florida gardening as if they're trying to win a fight. They see the heat, the humidity, the afternoon storms, the sand, and they decide they're going to force plants to grow despite all of it.

They amend harder. Water more. Fertilize heavier. Try to control every variable.

They lose. Every time.

Florida doesn't respond to control. It responds to observations.

The gardeners who succeed here stop fighting and start paying attention instead.

Perennials Succeed Because They Adapt, Not Because You Control Them

Annuals require you to recreate the same perfect conditions every single season. You plant it. You water on a schedule. On schedule, you fertilize. You harvest. Then you start over.

When conditions aren't perfect, the plant fails. If you miss a watering, it dies. If a storm floods the bed, you lose everything.

Annuals are high-input, high-maintenance, and high-risk in Florida.

Perennials are the opposite.

Perennials establish once, then adapt. They send deep roots that locate water sources. They determine when to grow and when to slow down. Their survival of storms, droughts, and heat waves.

You're not managing perennials. You're giving them the conditions they need to manage themselves.

That's a completely different approach.

Stop Following Schedules. Start Reading Reality.

I used to water every morning at 7 AM like clockwork. I thought I was being disciplined.

Overnight rain had already soaked the soil on some mornings. I watered anyway because it was 7 AM.

Some mornings the soil was bone dry because storms had skipped my yard for a week. I watered the same amount because it was 7 AM.

Then one morning I walked out and found my moringa with roots sitting in mud, leaves yellowing, stem soft at the base. Root rot.

Because I'd been watering on schedule instead of checking the soil first.

I disciplined myself. I was also killing plants.

Florida doesn't follow patterns. One week it rained two inches every afternoon. The next week it doesn't rain for ten days. One month the temperature stays in the 90s. The next month it drops into the 60s overnight.

Routines that worked in other climates fail here because Florida doesn't cooperate with routines.

Successful Florida gardeners don't follow schedules. They stroll through their garden. Instead of using a timer, they check soil moisture with their hands. They watch how plants respond to rain, heat, and storms.

They adjust in real time based on what's actually happening, not what a calendar says should be happening.

The moment I stopped watering on schedule and started watering based on what the soil and plants were telling me, I stopped losing plants to root rot and drought stress.

Not because I got better at gardening. Because I finally started paying attention.

Let the plant tell you what it needs

Plants communicate. Not with words. With signals.

Leaves drooping at 10 AM before the heat peaks? That's heat stress. The plant is reducing surface area to conserve water. Adding more water doesn't help. It needs shade or airflow.

Leaves drooping at 5 PM after a week of no rain? That's drought. Water now.

Yellowing leaves with green veins? Iron deficiency. High pH soil is locking it out.

Every problem the plant has shows up visually before it becomes fatal. But most people don't look. They follow advice from a book written for a different climate and wonder why it doesn't work.

I killed dozens of plants before I learned to read them.

I'd see yellowing leaves and assume nutrient deficiency. I'd add more fertilizer. The plant would get worse.

Because the problem wasn't a lack of nutrients. It was pH locking them out. More fertilizer just made the lockout worse.

The plant had been telling me the whole time. I wasn't listening.

My wife figured this out faster than I did. She'd walk through the beds and say things like, "That one's thirsty" or "That one's burning up, not drying out." I'd ask how she knew. She'd just shrug. "Look at it."

She was right. The plants show you everything if you pay attention.

You're not creating a garden. You're creating a system.

This is the shift that changed everything for me.

A garden is something you maintain. It is you who plants it. You give it water. You weed it. You harvest it. Then you do it all again next season.

A system is something that maintains itself.

Perennials are the foundation of a system. Once established, they grow without replanting. They spread. They shade the soil. They drop leaves that mulch themselves. These attract beneficial insects, pollinators, and predators that manage pests.

They create microclimates that allow other plants to thrive.

You're not gardening. You're building an ecosystem that produces food with less input every year.

Year one, you plant. Mulch. Water. You manage.

The plants established themselves in year two. You need to water less. Because plants shade the soil and drop their own organic matter, you mulch less. You manage less because the system is regulating itself.

Year three and beyond, the system runs. You reduced your spending, harvest, and add mulch occasionally.

But you're not fighting to keep things alive. The plants are thriving because the system supports them.

That's what perennial gardening in Florida looks like when it works. Not constant intervention. Observation and change.

The system does the work. You guide it.

Success in Florida means letting go of perfection

You will not have a pristine, weed-free, perfectly manicured garden in Florida.

If that's your goal, you'll be miserable.

Florida is messy. Things grow fast. Weeds show up. Vines climb where you don't want them to . Storms knock things over. Heat stresses plants. Pests move in.

And that's fine.

Because a messy garden that produces food year-round is better than a perfect garden that dies every summer.

Part of our intentionally messy garden. Thirty-plus edible crops in a small backyard space; this is what actual food production looks like, not a magazine spread.

I used to obsess over making everything look perfect. Neat rows. Weed-free beds. Uniform spacing.

It looked great for about two weeks.

Then a summer storm rolled through, flattened half the bed, and sent vines scrambling over everything I'd carefully staked. I spent the next month trying to restore order.

Then I let go.

I stopped pulling every weed. I let vines grow where they wanted as long as they weren't choking out productive plants. I stopped stressing over symmetry and started focusing on production.

And my garden got better. Not prettier. Better.

Because I was working with Florida instead of trying to make it look like a magazine spread.

Perfection isn't the goal. Production is.

One of my kids asked me once why the neighbors' yards looked neater than ours. I told him their yards don't feed us. Ours does.

He got it.

The plants in the next section work because they match Florida

The plants you're about to learn thrive here because they evolved from the heat, humidity, and poor soil. You don't need to recreate temperate conditions for them. They don't need constant attention. They don't need perfect soil.

They just need you to stop fighting Florida and start working with it.

Plant them in the right spot. Give them the drainage they need. Mulch them. Then let them adapt.

That's the Florida perennial mindset.

Skip the next section, and everything you just learned becomes theory. Apply it with the right plants, and you'll have a garden that produces year-round with less work each season.

Let's talk about the plants that actually work.

Chapter 4 The Heat Seekers

FULL SUN PERENNIALS THAT THRIVE IN HUMIDITY AND EXTREME HEAT. PLANTS THAT DON'T JUST SURVIVE FLORIDA SUMMERS BUT LOVE THEM.

M ost gardening advice tells you to give plants full sun. Six to eight hours minimum. More is better.

In Florida, full sun is not a gift. It is a test.

Full sun here is not the same as full sun in Colorado or Oregon or Pennsylvania. Full sun in Florida is 95 degrees with 80 percent humidity and no cloud cover for six hours straight. It is relentless. It is brutal. And most plants that claim to love full sun everywhere else will wilt, scorch, and struggle here.

But some plants thrive in it.

These are the heat seekers. Perennials that do not just tolerate Florida summers. They really like them. In warmer temperatures, they grow more quickly. Full sun leads to greater production. These conditions, which kill everything else, don't affect them.

These plants grew in tropical climates with extreme heat, high humidity, and poor soil. Florida is not a challenge for them. It is home. They require minimal water once established, tolerate neglect, and produce food year-round in most of the state.

This chapter is about those plants. But first, we need to talk about the one plant everyone kills first.

The Plant Everyone Kills First: Hibiscus

Hibiscus looks perfect for Florida. Tropical flowers. Loves heat. Handles humidity. Every nursery sells them. Every yard has them. And most people kill them within two years.

Roselle (Hibiscus saödariffa)

Not because hibiscus cannot handle Florida. Because people assume tropical means indestructible and forget it still needs proper drainage and pest management.

Hibiscus attracts every pest in Florida. Aphids, whiteflies, spider mites, hibiscus bud weevils. Add poor drainage and you get root rot. Add inconsistent watering and you get stress that invites more pests.

Hibiscus works in Florida, but only if you manage it. Just because a plant is tropical does not mean it is foolproof.

The plants in this chapter actually are.

If You Only Plant One: Start With Moringa

Moringa tree

If you have time, money, or space for only one heat-seeking perennial, plant moringa.

Moringa grows faster than anything else in this book. It yields more edible biomass per square foot than any other perennial. Even with extreme heat, poor soil, and neglect, it thrives. It grows year-round in South Florida, and it regrows from the roots after Central Florida freezes.

Every part of the plant is edible. Leaves taste like mild spinach. Pods taste like asparagus when young. Flowers are edible. Seeds produce oil.

Plant a cutting in spring and it will be six feet tall by summer. Prune it hard every few months and it produces leaves constantly. This is the one plant that justifies its space in every Florida yard.

If you plant nothing else from this chapter, plant moringa.

Moringa: The Fastest Growing Edible in Florida

Moringa oleifera is the fastest growing edible perennial in Florida.

Plant a cutting in spring and it will be six feet tall by summer. Ten feet by fall. Fifteen feet by the end of the year if you let it. It grows so fast you have to prune it constantly or it becomes a tree.

Moringa loves full sun. The hotter, the better. It grows year-round in South Florida. In Central Florida, it dies back in freezes but regrows from the roots in spring. In North Florida, treat it as an annual or grow it in a container you can move indoors during freezes.

How to Grow It:

Plant moringa in full sun with excellent drainage. Water regularly until established, then back off. Over-watering kills moringa faster than drought.

Fertilize lightly. Moringa grows so fast that heavy fertilization creates weak, floppy growth.

Prune it hard. Cut it back to four to five feet every few months to keep it bushy and productive. The more you prune, the more leaves it produces.

Harvest leaves constantly. Pick the entire leaflet stem, not individual leaves. Fresh growth comes back in days.

Why It Works:

Moringa evolved in tropical climates with poor soil and extreme heat. Florida matches its native environment perfectly. It produces more edible biomass per square foot than almost any other perennial in this book.

Cassava: Food Security, Not a Side Dish

My first time harvesting cassava from our garden.

Cassava is not a leafy green. It is a carbohydrate perennial that produces starchy tubers weighing five to ten pounds each.

This is food security. Not a garnish. Not a supplement. Calories.

Manihot esculenta produces more carbohydrates per square foot than potatoes and requires almost no maintenance. The leaves are also edible when cooked, though they require proper preparation to remove toxins.

Cassava loves full sun and heat. It grows year-round in South Florida. In Central Florida, it grows during warm months and slows in winter. In North Florida, it struggles and is better grown as an annual.

How to Grow It:

Plant cassava from stem cuttings, not seeds. Cut a mature stem into six to eight-inch sections and plant them at an angle in well-draining soil. They root in weeks.

Water regularly until established. Once established, cassava is drought tolerant.

Fertilize lightly. Too much nitrogen creates lush foliage and small roots. Cassava grows best in poor soil with minimal fertilization.

Harvest roots after eight to twelve months. Dig carefully around the base and pull the entire plant. You must cook the roots before eating them to remove cyanogenic glycosides.

Why It Works:

Cassava evolved in tropical South America. It thrives in heat, humidity, and poor soil. If you want a staple crop that grows year-round in Florida and actually feeds people, cassava is the answer.

Longevity Spinach: You Cannot Kill This Plant

Longevity spinach

Longevity spinach is the most indestructible perennial in this book.

Gynura procumbens handles full sun, partial shade, poor soil, drought, floods, and neglect. It grows as a low-spreading groundcover or a sprawling vine depending on how you manage it. It produces thick, succulent leaves with a mild, slightly bitter flavor.

I have seen longevity spinach survive weeks of drought, flooding rains, full sun, and zero fertilizer. It does not care. It grows anyway.

How to Grow It:

Plant longevity spinach from cuttings. Stick a stem in the ground and it roots in days.

Water occasionally until established. Once established, it handles drought and flooding equally well.

Fertilize rarely. Longevity spinach grows in pure sand with no amendments.

Harvest leaves anytime. Pick the top few leaves from each stem. Fresh growth comes back immediately.

Why It Works:

Longevity spinach is native to Southeast Asia, where it grows wild in ditches, roadsides, and disturbed areas. It evolved to handle extremes. Florida cannot stress it. This is the plant you give to someone who thinks they have a black thumb.

Katuk: The Rare Shade-Tolerant Heat Lover

Katuk one of my favorite perennials.

Katuk is on this list because it tolerates full sun in Florida, yet it actually favors partial shade. It is one of the few edibles that thrives in both.

Sauropus androgynus produces mild, slightly nutty leaves that taste like a cross between peas and spinach. It produces year-round in South and Central Florida and grows into a dense, bushy shrub. It prefers hot, humid conditions. Pest problems are rare.

How to Grow It:

Plant katuk in full sun to partial shade. It tolerates full sun but grows more tender leaves in partial shade.

Water regularly until established. Once established, katuk handles dry spells well but grows better with consistent moisture.

Prune regularly to keep it bushy. Katuk grows tall and sparse if you do not prune. Cut it back by half every few months. Harvest leaves from the top six inches of each branch.

Why It Works:

Katuk is native to Southeast Asia, where heat and humidity are constant. It matches Florida's summer conditions perfectly and tolerates shade better than almost any other edible perennial.

Okinawa Spinach: The Purple Groundcover That Thrives on Neglect

Okinawa spinach is like longevity spinach but with striking purple undersides on the leaves and a slightly distinct flavor.

Gynura crepioides produces thick, fleshy leaves with a mild, slightly tangy flavor. The tops are green. The undersides are deep purple. It can grow as a climbing vine or a sprawling groundcover. Full sun and heat are ideal for it. It grows faster in summer than any other time of year.

How to Grow It:

Plant Okinawa spinach from cuttings. Stick a stem in the ground and it roots in days.

Water occasionally until established. Once established, it handles dry spells and wet spells equally well.

Harvest leaves anytime. Pick the top few leaves from each stem. Fresh growth comes back immediately.

Why It Works:

Okinawa spinach is native to Southeast Asia. It thrives in heat and humidity. If you want a perennial green with striking color and zero maintenance, this is it.

Brazilian Spinach: The Climbing Perennial That Grows Anywhere

Brazilian spinach is not a true spinach, but it produces edible leaves that taste mild and slightly mucilaginous, similar to okra.

Alternanthera sissoo grows as a sprawling groundcover or a climbing vine. It loves full sun and heat. It produces year-round in Central and South Florida. It handles drought and poor soil without issue. Cuttings root in days.

How to Grow It:

Plant Brazilian spinach from cuttings. It roots easily in soil or water.

Water occasionally until established. Once established, it handles drought well but grows faster with consistent moisture.

Harvest leaves anytime. Pick the top few inches of each stem. Fresh growth comes back in days.

Why It Works:

Brazilian spinach evolved in tropical South America. It thrives in heat, humidity, and poor soil. If you want a perennial green that climbs or spreads and requires zero maintenance, this is it.

What These Plants Have in Common

Every plant in this chapter evolved in tropical climates with extreme heat, high humidity, and poor soil. Florida is not a challenge for them. It is their natural environment.

Full sun is what they love. Faster growth occurs in heat for them. They manage humidity well, avoiding fungal issues. Minimal water is required by them once established. They produce year-round in most of Florida.

These are not temperamental plants that need perfect conditions. These are tough, adaptable, low-maintenance perennials that turn Florida's extreme heat into an advantage.

Plant them in full sun. Give them decent drainage. Mulch them once. Then step back and let them grow.

But what about the rest of your yard?

If you stop here, you will still fail in half your yard.

Most Florida properties have shade. Live oaks, palms, structures, and other plants. Full sun perennials die there. The plants in this chapter will not save those spaces.

The plants in the next chapter will.

Most edibles fail in the shade. The perennials you are about to learn do not just tolerate shade. It is a place where they thrive. They produce more in partial shade than they do in full sun. They solve the problem every Florida gardener faces: what to plant under the trees.

Skip the next chapter, and half your yard stays empty. Read it and you'll turn every shaded space into production.

Chapter 5 The Shade Survivors

PERENNIALS THAT THRIVE UNDER LIVE OAKS, PALMS, AND PARTIAL SHADE. REAL SOLUTIONS FOR REAL FLORIDA YARDS, NOT FANTASY GARDEN BEDS.

M ost Florida yards have shade.

Live oaks. Palms. Structures. Other plants. The shade isn't optional. It's permanent.

And most gardening advice ignores it entirely.

Books tell you to plant in full sun. Videos tell you edibles need six to eight hours of direct light. Seed packets tell you nothing will produce in shade.

And you look at your yard and realize half of it is under trees. The other half gets blasted by an afternoon sun that scorches everything.

Shade provides the only comfortable spots. And nothing you plant there survives.

A large oak tree shading a third of the north end of my backyard garden.

This is the Florida gardening problem nobody talks about. Shade isn't the exception. It's the rule.

And most edibles fail in it.

The plants in this chapter don't.

These are perennials that thrive in partial shade. They produce more tender leaves in shade compared to full sun. Underneath tree canopies, they experience faster growth. Areas that receive filtered light, dappled shade, or just morning sun are within their capabilities.

These aren't shade-tolerant plants that barely survive. These are shade-preferring plants.

But first, we need to talk about the plant everyone kills in the shade.

The Plant Everyone Kills in Shade: Basil

Basil looks like it should work in Florida shade. It wilts in full sun. It needs protection from heat.

So people plant it under trees, thinking the shade will help.

It doesn't.

Basil in shade stretches tall and leggy, reaching for light it can't find. Small leaves remain, and the stems grow weak. Harvesting is never possible because of insufficient plant yield.

Then humidity moves in, and fungal diseases finish it.

Basil needs full sun in temperate climates and bright, indirect light in Florida. No shade.

Partial shade under a tree isn't bright indirect light. It's low light with high humidity. That combination kills basil every time.

I tried this twice in my backyard before I figured it out. Planted basil under the live oak, thinking I was being smart. Protecting it from the afternoon sun. Giving it the shade it needed.

Both plants stretched four feet tall trying to reach light, never got bushy, and died from fungal diseases within two months.

The problem isn't shade. The problem is planting crops that need high light levels in areas that can't provide them.

The plants in this chapter actually prefer shade. They produce more in it. No stretching toward light. No struggling.

They thrive.

Why Shade Wins in Florida?

Before we get to specific plants, you need to understand why shade is an advantage in Florida, not a limitation.

Less Heat Stress

Full sun here isn't just light. It's 95 degrees with 80% humidity and no cloud cover for six hours straight.

That level of intensity stresses most plants. They close their stomata to conserve water. Growth slows. Production drops.

Partial shade moderates temperature. Plants stay cooler. Stomata stay open longer. Photosynthesis continues without the plant fighting to survive.

More Consistent Moisture

Full sun dries out soil fast. Water evaporates before roots can use it. You water more. The plant still stresses.

Partial shade slows evaporation. The soil stays moist longer. Roots access water consistently.

The plant grows steadier.

More tender leaves, fewer pests

Plants grown in full sun develop thick, waxy leaves to prevent water loss. That makes them tougher and less palatable.

Plants grown in partial shade produce thinner, more tender leaves. They taste better. They cook faster.

When the plant remains unstressed, the pest pressure decreases.

Keep these three reasons in mind as you read each plant profile. Shade isn't a problem to solve. It's a condition to leverage.

If you only plant two for shade

If you have time, money, or space for only two shade perennials, here's how to choose:

For set-it-and-forget-it production: Longevity spinach and Okinawa spinach. They are both indestructible. Both achieve year-round production. They both spread on their own.

For wet shade areas: Sisso spinach and sweet potato leaves. Both handle standing water and occasional flooding.

For edible beauty: Cranberry hibiscus and katuk. Both produce food and look striking in the landscape.

For maximum yield: Chaya and katuk. Both produce more edible biomass per square foot than any other shade perennial.

Now let's get to the plants.

Katuk: The Daily Greens Champion

We grow our katuk in partial shade on the north end of our backyard and it does fine.

Katuk appeared in Chapter 4 because it handles full sun. But shade is where it shines.

Sauropus androgynus produces mild, slightly nutty leaves that taste like a cross between peas and spinach.

In full sun, it grows. In partial shade, it thrives.

I planted katuk in full sun first. It grew fine. Then I moved one to partial shade under a live oak. That plant outproduced the full-sun plant by double within six months.

Softer were the leaves, and the flavor was better. Instead of getting leggy, the plant maintained its compact form.

My wife noticed. She kept harvesting from the shaded plant and ignoring the sun-grown one. When I asked why, she said, "The leaves from this one actually taste good."

She was right.

Role: Best for daily greens. Mild flavor, year-round production, works raw or cooked.

How you'll use it: Salads, smoothies, stir-fries, wraps, spring rolls.

How to grow it:

Plant katuk in partial shade where it gets morning sun or filtered light through tree canopies. Water regularly until established. Fertilize lightly every few months.

Prune regularly to keep it bushy.

How to harvest: Pick leaves from the top six inches of each branch. Don't strip the entire plant. Leave lower leaves intact. The plant regrows from what you leave behind.

Why it works:

Katuk grew in the understory of Southeast Asian forests. Partial shade mimics its natural environment. The plant produces more with less stress.

Chaya: The High-Yield Powerhouse

Chaya is the most productive perennial green in Florida. And it prefers shade.

Cnidoscolus aconitifolius produces large, lobed leaves that taste like mild spinach when cooked. It grows into a large shrub or a small tree. It produces year-round in South and Central Florida.

Cook the leaves. Never eat them raw.

Raw leaves contain toxins that cooking destroys.

I planted chaya in full sun first. It grew slowly. The leaves scorched in summer.

Then I moved one to partial shade under a palm. That plant exploded.

It grew twice as fast. The leaves stayed dark green. Production tripled.

Role: Best for high-yield production. Cooks down like spinach, producing massive quantities.

How you'll use it: Soups, stir-fries, sauteed greens, anywhere you'd use spinach or collards.

How to grow it:

Plant chaya in partial shade with decent drainage. Water occasionally until established. Fertilize lightly.

Prune regularly. Chaya grows tall and leggy if you don't prune it. Cut it back by half every few months.

How to harvest: Cut stems with leaves from the top six inches. Boil or sauté before eating. Never eat raw.

Why it works:

Chaya is native to the understory of Central American forests. It evolved for shade. In Florida, partial shade reduces heat stress and increases production dramatically.

Longevity Spinach: Set It and Forget It

Longevity spinach appeared in Chapter 4 as a heat seeker. It also thrives in partial shade.

Better, actually.

Gynura procumbens handles full sun, partial shade, poor soil, drought, floods, and neglect.

In full sun, it grows fast. In partial shade, it grows even better. The leaves are more tender. The plant spreads faster.

*Recently planted longevity spinach in a 15g grow
bag - north end, partial shade. It'll thrive here.*

I have longevity spinach in full sun and partial shade in my backyard. The shaded plants outproduce the full-sun plants consistently.

The leaves are softer. The flavor is milder.

Role: Best for beginners and people who forget to water. Indestructible.

How you'll use it: Salads, smoothies, teas, stir-fries.

How to grow it:

Plant longevity spinach from cuttings in partial shade. Stick a stem in the ground. It roots in days.

Water occasionally until established. Fertilize rarely.

How to harvest: Pick leaves anytime. The plant regrows constantly.

Why it works:

Longevity spinach is native to disturbed areas and forest edges in Southeast Asia. It evolved to handle extremes in low light.

Florida shade can't stress it.

Cranberry Hibiscus: The Edible Ornamental

Cranberry hibiscus is one of the few edibles that works as both food and a striking landscape plant.

Hibiscus acetosella produces deep red-purple leaves with a tart, cranberry-like flavor. Young leaves are edible raw in salads. Older leaves are better cooked.

The plant grows into a dense shrub with a color that stops people in their tracks.

I planted cranberry hibiscus in my front yard in partial shade. Neighbors keep asking what it is. They assume it's ornamental. Then I tell them I eat it.

Role: Best edible ornamental. Beautiful and functional.

How you'll use it: Young leaves in salads, older leaves cooked like spinach, added to soups for color and tang.

How to grow it:

Plant cranberry hibiscus in partial shade with decent drainage. Water regularly until established.

Fertilize lightly. Too much nitrogen reduces the red color.

Prune regularly to keep it bushy.

How to harvest: Pick young leaves from the top of the plant. Use them fresh or cook them. The plant regrows constantly.

Why it works:

Cranberry hibiscus is native to tropical Africa, where it grows in forest edges with partial shade. It handles Florida heat and humidity but produces better leaves and more intense color in shade.

Okinawa Spinach: The Purple Groundcover

Okinawa spinach also appeared in Chapter 4. Like longevity spinach, it grows better in partial shade than full sun.

Gynura crepioides produces thick, fleshy leaves with a mild, slightly tangy flavor. The tops are green. The undersides are deep purple.

It grows as a sprawling ground cover or a climbing vine.

In full sun, Okinawa spinach grows fast but requires more water. In partial shade, it thrives with less input. The leaves remain more tender. The purple color intensifies.

Role: Best for ground cover. Spreads fast, striking color, zero maintenance.

How you'll use it: Salads, stir-fries, wraps.

How to grow it:

Plant Okinawa spinach from cuttings in partial shade. Water occasionally until established.

How to harvest: Pick leaves anytime. The plant regrows from every node.

Why it works:

Okinawa spinach is native to forest edges in Southeast Asia. Shade reduces heat stress and increases both production and color intensity.

Sisso Spinach: The Wet Shade Solution

Sisso spinach solves the problem nobody else addresses. Wet shade.

Alternanthera sessilis produces small, mild-flavored leaves that grow on sprawling stems. It works as a ground cover or a trailing plant.

It loves partial shade and handles wet soil better than almost any other edible.

I planted sisso spinach in a low spot that floods after every rain. Full-sun plants died there within weeks.

Sisso spinach thrived.

It spread across the entire area and produced edible leaves constantly.

Role: Best for wet shade. The only edible that actually wants standing water.

How you'll use it: Stir-fries, soups, salads.

How to grow it:

Plant sisso spinach from cuttings in partial shade. Water regularly until established. Once established, it handles wet soil and occasional flooding.

Fertilize rarely.

How to harvest: Cut stems with leaves. The plant regrows from every node.

Why it works:

Sisso spinach is native to wetland edges and disturbed areas in tropical Asia. It evolved for shade and wet soil.

In Florida, it thrives in the spots where everything else drowns.

Sweet Potato Leaves: The Dual-Purpose Plant

Sweet potato bed we created for a dual harvest.
You can eat the leaves like spinach while you wait
for the root vegetable to become ready.

People usually grow sweet potatoes in full sun for tubers. But the vines tolerate partial shade better than most people realize.

Ipomoea batatas produces edible leaves and tubers.

In full sun, the plant focuses energy on tuber production. In partial shade, it produces more leaves and fewer tubers.

If you're growing sweet potatoes for greens, not tubers, partial shade works perfectly.

The leaves are mild, slightly sweet, and highly nutritious. They produce year-round in South and Central Florida.

Role: Best dual-purpose plant. Leaves in shade, occasional tubers as a bonus.

How you'll use it: Sautéed greens, soups, stir-fries.

How to grow it:

Plant sweet potato slips in partial shade with decent drainage. Water regularly until established. Fertilize lightly.

How to harvest: Pick leaves from the top six inches of the vine. Leave the lower vines intact.

Why it works:

Sweet potato vines tolerate shade better than most root crops. If you want edible leaves and occasional tubers in shaded areas, this is it.

Your shaded areas aren't a problem anymore

Walk outside today. Look at the spots under your trees. Under your palms. Next to structures.

Those aren't dead zones. Those are production zones.

The plants in this chapter turn shade into an advantage. They experience more rapid growth and produce more tender leaves. They handle neglect better because they're not fighting extreme heat.

Plant them under live oaks, palms, or structures. Give them decent drainage. Mulch them once. Then step back and let them grow.

But Florida yards aren't just divided into sun and shade. Some areas face challenges that go beyond light levels.

Coastal properties deal with salt spray that kills most edibles. Inland areas face water restrictions. Low-lying spots flood after every storm. High ground dries out in days.

Most perennials fail under those conditions.

The plants in the next chapter don't. They handle salt, drought, flooding, and neglect. They're the toughest perennials in this book.

Skip the next chapter and you'll still fail in problem areas. Read it and you'll turn every tough spot in your yard into production.

Chapter 6 Native Perennials That Feed the Ecosystem

FLORIDA NATIVES THAT SUPPORT POLLINATORS, BIRDS, AND BENEFICIAL INSECTS. PRACTICAL, BEAUTIFUL, AND LOW MAINTENANCE WHEN PLANTED CORRECTLY.

M ost of this book focuses on edible perennials from other tropical regions. Moringa from India. Katuk from Southeast Asia. Cassava is from South America.

Those plants work in Florida because Florida's climate matches their native environments. They're tough, productive, and low maintenance once established.

But here's what surprised me after ten years of gardening in west-central Florida.

The toughest, lowest-maintenance plants in my backyard aren't always the tropicals. They're the natives.

Natives aren't charity plants you grow out of obligation. They're high-performance plants that repair the ecosystem at the same time.

Less water is required by them once established. They handle pests better because they co-evolved with the insects that live here. Florida soil is all they need to thrive, so fertilizer is unnecessary.

And many of them produce food.

This chapter is about Florida native perennials that feed you, feed wildlife, and require almost no maintenance once established.

These are functional natives, not ornamentals you plant for looks and never touch.

But before we get to the plants that work, we need to talk about the native everyone plants wrong.

The Native Everyone Plants Wrong: Elderberry in Dry Soil

Fresh elderberries from one harvest off two trees nearly 15 lbs total. ONE harvest!

Elderberry is native to Florida. It produces clusters of dark purple berries that are excellent when cooked or made into syrup.

Birds love it. People love the idea of growing native berries.

And most people kill it by planting it in dry soil.

Elderberry evolved in wet areas. Creek banks. Drainage swales. Low spots that stay moist year-round.

It doesn't tolerate dry soil.

Plant it on high ground or in sand that drains fast, and it struggles. The leaves scorch. Growth slows. Production drops.

I planted elderberries in a raised bed the first time. I watered it regularly. It never thrived. I lacked experience in growing things and believed I could keep elderberry plants small.

I'm not sure what I was thinking.

Then I moved one to a spot near my property line that floods after every rain.

That plant exploded.

It grew three times as fast. It produced berries heavily. Because I finally put it where it belonged.

Elderberry works in Florida, but only in wet soil.

Native doesn't mean it grows anywhere. It means it's adapted to specific conditions.

Match the plant to the right spot, and it thrives. Ignore that and it struggles like anything else.

The plants in this chapter come with clear guidelines on where they actually belong.

Why Native Plants Are Easier, Not Harder

Before we get to specific plants, you need to understand why planting natives is an upgrade, not a sacrifice.

Natives Require Less Water

Native plants evolved in Florida with Florida's rainfall patterns. They don't need irrigation systems once established. They survive dry spells because they have adapted to them.

In my backyard, the natives go months without supplemental water. The non-natives need watering weekly during dry spells.

That's less work and lower water bills.

Natives Handle Pests Better

Non-native plants often attract pests that have no natural predators here. Native plants co-evolved with native insects, which means natural predator-prey relationships keep pest populations in check.

I have passionflower vines that get eaten by caterpillars constantly. That's the point.

Butterflies emerge from caterpillars. The plant regrows. The system regulates itself.

I spray nothing.

Natives Need Zero Fertilizer

Native plants evolved in Florida's soil. They do not need amendments. They don't need fertilizer. These grow within the conditions that are already present.

My coontie has been in pure sand for five years. I've never fertilized it. It looks perfect.

Because it evolved to grow in pure sand.

Natives Support Pollinators That Help Everything Else

Many native bees, butterflies, and other pollinators evolved alongside specific native plants.

When you plant natives, you're feeding the insects that pollinate your vegetables, your fruit trees, and everything else in your yard.

My wife and I started planting more natives three years ago. We noticed an immediate increase in native bees and butterflies.

Our vegetable garden production went up because pollination improved.

The natives weren't just growing food themselves. They were improving the productivity of everything around them.

If you only plant two natives

If you have time, money, or space for only two native perennials, here's how to choose:

For maximum ecosystem support: Coontie and passionflower. Coontie supports the rare Atala butterfly. Passionflower supports Gulf Fritillary and Zebra Longwing butterflies. Both require zero maintenance.

Passion fruit flowers are easily one of the most stunning blooms in my entire garden. The fruit comes later, but this view alone is worth growing them.

For edible production: Saw palmetto and elderberry. Both produce berries with zero fertilizer, zero pesticides, and minimal water.

For extreme drought tolerance: Coontie and yaupon holly. Both handle pure sand, no irrigation, and zero fertilizer. Perfect for properties with water restrictions.

For wet areas: Elderberry and pickerelweed. Both thrive in standing water where most edibles drown.

Now let's get to the plants.

Coontie: The Indestructible Butterfly Magnet

Coontie is one of the toughest native plants in Florida. And it supports a butterfly that exists nowhere else.

Zamia integrifolia is a cycad that looks like a small palm or fern. It grows slowly but lives for decades. It handles full sun, partial shade, drought, poor soil, and neglect.

It's native to all of Florida.

Coontie is the only host plant for the Atala butterfly, which was nearly extinct until people started planting coontie again.

I planted three coonties in my backyard four years ago. Within six months, I saw my first Atala butterfly.

Now I see them constantly. Their metallic blue wings are unmistakable.

One of my kids spotted the first one. Ran inside yelling about the "robot butterfly." That's what they look like—metallic blue and black, almost iridescent.

Historically, indigenous peoples and early settlers processed coontie roots into flour. The roots have toxins that require leaching before consumption, making it unsuitable as a casual harvest crop.

Grow coontie for the butterflies, not the roots.

Role: Best native for supporting rare pollinators and extreme drought tolerance.

How to grow it:

Plant coontie in full sun to partial shade with excellent drainage. Water occasionally until established. Once established, coontie is drought tolerant and requires zero maintenance.

Don't fertilize. Coontie grows in pure sand with no amendments.

Why it works:

Coontie evolved in Florida scrub habitats with poor soil and low rainfall. It's one of the most drought-tolerant native plants in the state.

If you want a native that requires zero input and supports rare wildlife, this is it.

Saw Palmetto: The Edible Berry That Grows Anywhere

Saw palmetto is native to all of Florida. People have used the edible berries it produces medicinally for centuries.

Serenoa repens grows as a low, sprawling shrub with sharp, saw-toothed leaves. It handles full sun, partial shade, drought, poor soil, flooding, and fire.

It's one of the most adaptable native plants in the state.

Berries ripen in fall, and people can eat them raw or dried. They have a sweet, slightly medicinal flavor.

Birds and wildlife also rely on them as a food source.

On GrowFitFL, I've gotten more questions about saw palmetto than almost any other native. People want to know if it's worth growing.

Yes. It is.

It requires zero maintenance and produces berries every year.

Role: Best native for edible berries and extreme adaptability.

How to grow it:

Plant saw palmetto in full sun to partial shade. It tolerates wet soil and dry soil equally well. Water occasionally until established. Once established, saw palmetto requires zero maintenance.

Don't fertilize.

Harvest berries in the fall when they turn dark purple or black.

Why it works:

Saw palmetto evolved in Florida's fire-adapted ecosystems. It handles drought, flooding, and poor soil because it had to.

It produces berries that support wildlife and humans. It's one of the lowest-maintenance natives in this book.

Elderberry: The High-Yield Berry for Wet Areas

Elderberry is native to North and Central Florida. It produces clusters of dark purple berries that are excellent cooked or made into syrup, jam, or wine.

Sambucus canadensis grows as a large shrub or small tree. It prefers wet soil and partial shade, but tolerates full sun. It produces white flowers in spring that attract pollinators, followed by berries in summer.

Cook the berries. Don't eat the berries raw.

Raw elderberries contain toxins that cause stomach upset. The flowers are edible and often used to make tea or fritters.

As I mentioned earlier, I killed my first elderberry by planting it in dry soil. Once I moved it to a wet spot that has some wind protection, it thrived.

If you have a drainage area, a low spot, or anywhere that stays moist, elderberry belongs there.

Role: Best native for high-yield berry production in wet areas.

How to grow it:

Plant elderberry in partial shade to full sun with consistent moisture. Elderberry prefers wet soil and grows well near ponds, drainage areas, or low spots.

Water regularly until established. Fertilize lightly if growth is slow.

Prune elderberry heavily every winter to keep it productive. Old canes produce fewer berries.

Harvest berries in summer when they turn dark purple.

Why it works:

Elderberry is native to wet areas in North and Central Florida. It thrives in spots where most edibles drown.

It produces heavily and supports pollinators.

If you have wet soil and want native berries, elderberry is the answer.

Yaupon Holly: The Native Caffeinated Tea

Yaupon holly is the only native North American plant that contains caffeine.

Ilex vomitoria grows as a large shrub or a small tree. It's native to coastal areas and inland Florida. It handles salt spray, drought, poor soil, and neglect.

The leaves are used to make a tea that tastes similar to green tea or yerba mate.

Despite the species name, yaupon doesn't cause vomiting. The name comes from a misunderstanding of indigenous ceremonial practices.

People have consumed tea for thousands of years.

My wife and I started harvesting yaupon leaves two years ago. We dry them and steep them like tea.

It's mild, slightly earthy, and caffeinated.

We stopped buying tea bags.

Female plants produce bright red berries in fall and winter that attract birds.

Role: Best native for caffeinated tea and coastal tolerance.

How to grow it:

Plant yaupon holly in full sun to partial shade. It tolerates salt spray, making it excellent for coastal properties.

Water occasionally until established. Once established, yaupon is drought tolerant and requires zero maintenance.

Harvest young leaves anytime. Dry them and steep like tea.

Why it works:

Yaupon holly is native to coastal Florida and inland areas with poor soil. It evolved to handle salt, drought, and neglect.

It produces caffeine naturally, making it one of the unique native edibles in the state.

Passionflower: The Butterfly Host That Produces Fruit

Passionflower is native to Florida and produces both edible fruit and stunning flowers that attract butterflies.

Passiflora incarnata grows as a climbing vine. It produces intricate purple and white flowers followed by small, edible fruits.

The fruit is tart and seedy, but excellent in smoothies or made into juice.

Passionflower is the host plant for Gulf Fritillary and Zebra Longwing butterflies.

I have passionflowers growing on a fence in my backyard. The caterpillars strip the leaves regularly.

The plant regrows within days. The butterflies are constant.

Role: Best native for attracting butterflies and producing fruit.

How to grow it:

Plant passionflower in full sun to partial shade with decent drainage. It grows as a vigorous vine and needs support or a fence to climb.

Water regularly until established.

Expect caterpillars to eat the leaves. That's the point. The plant regrows quickly.

Harvest fruit when it turns from green to yellow or purple and falls from the vine. Eat fresh or juice.

Why it works:

Passionflower is native to disturbed areas, roadsides, and forest edges throughout Florida. It evolved to be eaten by caterpillars and regrow constantly.

It produces fruit and supports butterflies simultaneously.

Persimmon: The Fall Fruit Tree

American persimmon is native to North and Central Florida. It produces sweet, orange fruit in fall that tastes like honey when fully ripe.

Diospyros virginiana grows as a medium to large tree. It handles full sun, partial shade, drought, and poor soil.

The fruit is astringent when unripe but delicious when soft and fully ripe.

Wildlife loves persimmons. Deer, raccoons, opossums, and birds will compete with you for the fruit.

I have a persimmon tree on my property line. I harvest what I can in the morning. By afternoon, the wildlife has cleaned out the rest.

Role: Best native fruit tree for fall harvest and wildlife support.

How to grow it:

Plant persimmons in full sun with decent drainage. Water regularly for the first year. Once established, persimmons are drought tolerant and require zero maintenance.

Harvest fruit in fall when it turns deep orange and soft. If the fruit is astringent, it's not ripe yet. Wait until it's mushy.

Why it works:

Persimmon is native to upland areas in North and Central Florida. It evolved to handle drought and poor soil.

It produces heavily and supports wildlife.

Pickerelweed: The Edible Wetland Native

Pickerelweed is native to Florida wetlands. It produces edible seeds and young leaves.

Pontederia cordata grows in shallow water or wet soil. It produces blue-purple flower spikes in summer that attract pollinators.

You can eat the seeds raw or cooked, and they taste similar to nuts.

Role: Best native for wet areas and edible seeds.

How to grow it:

Plant pickerelweed in wet soil or shallow water. It thrives in pond edges, drainage areas, or low spots that stay wet.

Water isn't an issue. This plant wants to be wet.

Harvest seeds in late summer when the seed heads dry. Young leaves are also edible cooked.

Why it works:

Pickerelweed is native to Florida wetlands. It evolved to grow in standing water.

If you have a wet area where nothing else survives, pickerelweed thrives there and produces edible seeds.

Why This Matters?

When you plant natives, you're not just growing food. You're rebuilding habitat.

Florida has lost millions of acres of native plant communities to development and invasive species.

Every native plant you add creates a refuge for pollinators, birds, and beneficial insects.

You don't need to convert your entire yard. Even a few native plants make a difference.

The more natives you plant, the more you'll see butterflies, native bees, and birds. Your yard becomes part of a larger network of habitat.

These plants aren't charity projects

Every plant in this chapter produces food, supports wildlife, or both.

These aren't ornamental natives you plant out of obligation and never use.

These are functional natives that reduce your workload, lower your water bill, and improve your yard's ecology at the same time.

Plant them in the conditions they evolved. Give them decent drainage. Mulch them once. Then step back and let them grow.

But not every part of your yard is standard Florida habitat. Some areas face extreme challenges that even tough natives struggle with.

Coastal properties deal with salt spray that kills most plants within months. Inland areas face water restrictions that make irrigation impossible. Low-lying spots flood after every storm. High ground dries out in days.

Most perennials fail under those conditions.

The plants in the next chapter don't. They handle salt, drought, flooding, and the neglect that would kill everything else in this book.

Skip the next chapter and coastal areas, drought zones, and flood-prone spots stay barren.

Read it and you'll turn the toughest areas in your yard into production.

Chapter 7 Low Water and Coastal Perennials

SALT TOLERANT, DROUGHT RESISTANT PLANTS FOR BEACHSIDE YARDS AND WATER RESTRICTED AREAS. TOUGH PLANTS FOR TOUGH ENVIRONMENTS.

S ome Florida yards aren't just challenging. They're extreme.

Coastal properties deal with salt spray that kills most edibles within weeks. Inland properties face water restrictions that make irrigation systems useless. High ground in sandy areas dries out days after rain. Low-lying spots flood, then turn bone dry when the water drains away.

Most perennials fail in these conditions. Even the tough tropicals struggle.

The natives in the last chapter handle a lot, but salt spray and severe drought push many of them past their limits.

The plants in this chapter don't have limits. Or at least, their limits are so far beyond normal that they feel indestructible.

In my backyard, I have a section that gets blasted by afternoon sun, never gets watered, and sits in pure sand. Most plants I put there die within weeks.

The plants in this chapter have been there for three years with zero input. They look better now than when I planted them.

This chapter is about removing failure from the equation. These are plants you can't kill even if you try.

But first, we need to talk about the plant everyone thinks is tough enough for these conditions, but isn't.

The Plant That Looks Tough But Isn't: Aloe Vera

Aloe vera looks indestructible. What you're looking at is a succulent that stores water and handles heat.

People assume it thrives on neglect.

It doesn't.

Aloe vera handles drought fine. It handles heat fine. But it doesn't handle salt spray. It doesn't handle flooding followed by drought. And it doesn't handle full afternoon sun in Florida sand without occasional watering.

Aloe doing double duty in our garden. We use the gel to help root cuttings from other plants, and slap it on every sunburn and bug bite the kids get.

I planted aloe vera in a dry, sandy spot, thinking it would be perfect. It survived for six months.

Then a few heavy rains flooded the area. The aloe sat in standing water for two days. Root rot set in. The plant collapsed.

Cost me twelve dollars at the nursery. Lasted six months. Died in a puddle.

Aloe works in containers where you control drainage. It works in raised beds with excellent drainage.

But in problem areas with unpredictable moisture and salt exposure, aloe fails more often than people expect.

The plants in this chapter are actually indestructible.

Why These Plants Survive What Others Can't

These perennials grew in coastal areas, deserts, or regions with long dry seasons. They handle salt, drought, and poor soil because they had no choice.

In Florida, those same conditions are advantages. The plants do not experience stress. The environment helps the plants thrive because it was designed for them.

Some are succulents that store water in their leaves or stems. Others have deep taproots that access water far below the surface. A few go dormant during extreme drought and bounce back when rain returns.

Most plants die when salt builds up in their tissues. These plants either excrete salt, sequester it in old leaves, or tolerate it at concentrations that would kill other perennials.

That makes them perfect for coastal properties where salt spray is constant.

Once established, these plants survive on rainfall alone. They don't need fertilizer. They don't need amendments.

They grow in pure sand with no help from you.

That's not an exaggeration. That's what they do.

If you only plant two for extreme conditions

If you have time, money, or space for only two perennials for extreme environments, here's how to choose:

For coastal properties with salt spray: sea purslane and prickly pear cactus. Sea purslane grows on beaches and handles direct salt exposure. Prickly pear tolerates coastal conditions and produces edible pads and fruit.

For inland drought with no irrigation: Prickly pear cactus and agave. Both survive months without water in pure sand.

For areas that flood and then **dry out:** Sea purslane and saltbush. Both handle wet feet temporarily and bounce back when conditions dry out.

For maximum edible production in extreme conditions: prickly pear cactus and sea purslane. Both produce food with zero water and zero maintenance.

Now let's get to the plants.

Prickly Pear Cactus: The Anchor Plant for Extreme Drought

Prickly pear blooms are hummingbird magnets.
They get first dibs on every flower, which means less
fruit for me but way more entertainment.

Prickly pear cactus produces edible pads and fruit in conditions that kill everything else.

Opuntia species are native to arid regions but thrive in Florida's sandy, drought-prone areas. The pads, called nopales, are edible when young and taste similar to green beans. People eat the sweet fruit, called tunas, fresh or make it into juice.

I planted prickly pear in the driest, sandiest part of my yard four years ago. I watered it twice in the first month. After that, I stopped.

I've never watered it since. I've never fertilized it. It's tripled in size and produces fruit every year.

Last summer, we went three months without rain. Everything else in that section looked stressed. The prickly pear looked perfect.

It was getting bigger, flowering, and producing fruit.

While I was watering other parts of the yard, this plant was thriving on its own.

On GrowFitFL, people ask me all the time if prickly pear really needs zero water.

Yes. It does.

Once established, you can forget it exists, and it'll still produce.

Role: Best for extreme drought and edible production with zero input. This is the anchor plant for inland areas with no irrigation.

How you'll use it: Nopales in stir-fries, tacos, and salads. Fruit, fresh or in juice, agua fresca, jams.

How to grow it:

Plant prickly pear in full sun with excellent drainage. Handle pads with thick gloves. The obvious spines are bad, but the tiny glochids are worse.

Water occasionally for the first month. After that, stop. Prickly pear survives on rainfall alone.

Don't fertilize. Prickly pear grows in pure sand.

How to harvest: Burn off spines over a flame or peel them carefully. Slice young pads and cook like green beans. Harvest fruit when it turns red or purple. Peel carefully and eat fresh or juice.

Why it works:

Prickly pears grow in deserts where rain comes once or twice a year. It stores water in its pads and survives months without moisture.

In Florida, it grows faster than it does in arid climates because it actually gets occasional rain.

This is the most drought-tolerant edible in the book.

Sea Purslane: The Anchor Plant for Coastal Conditions

Sea purslane is the toughest edible perennial for coastal properties. It grows on beaches, handles direct salt spray, tolerates flooding, and survives drought.

Sesuvium portulacastrum produces thick, succulent leaves that are edible raw or cooked. The flavor is mild and slightly salty. It grows as a sprawling groundcover that spreads across sand.

I've seen sea purslane growing on dunes with nothing but pure beach sand and constant salt spray. No soil. No irrigation. No fertilizer.

Just sand, salt, and sun. And it thrives.

My wife and I planted sea purslane along the property line near the coast three years ago. It gets blasted with salt spray during storms. It floods when it pours. Then it dries out for weeks.

The plant doesn't care.

It keeps growing. It keeps producing leaves. We harvest from it constantly.

On GrowFitFL, people ask if sea purslane really grows in pure beach sand.

Yes. It does.

I've seen it. I've grown it. This is the most salt-tolerant edible in Florida.

Role: Best coastal edible and most salt-tolerant perennial. This is the anchor plant for beachside properties.

How you'll use it: Raw in salads, cooked like spinach, pickled, in smoothies.

How to grow it:

Plant sea purslane in full sun with excellent drainage. It tolerates pure sand, direct salt spray, and occasional flooding.

Water occasionally until established. Once established, it survives on rainfall alone.

Don't fertilize.

How to harvest: Pick leaves anytime. The plant regrows constantly.

Why it works:

Sea purslane grows on beaches. At the cellular level, it handles salt. It stores water in its leaves. It tolerates flooding and drought in the same week.

If you live near the coast, this plant thrives where nothing else survives.

Agave: The Set-It-and-Forget-It Succulent

Agave survives extreme drought with zero water and produces edible flowering stalks.

Several Agave species thrive in Florida, particularly in sandy, drought-prone areas. The flowering stalk, called a quiote, is edible when roasted.

Agave stores water in its thick leaves and can survive a year or more without rain.

Role: Best for extreme drought and ornamental value. Choose agave over prickly pear if you want a striking landscape appearance with edible flowers.

How you'll use it: Roasted flower stalks like asparagus.

How to grow it:

Plant agave in full sun with excellent drainage. Water occasionally for the first month, then stop.

Be cautious of sharp leaf tips. Some species have spines that can cause serious injury.

How to harvest: When the flowering stalk emerges (this can take years), cut it and roast it.

Why it works:

Agave grows in arid climates with months between rains. It stores water and shuts down growth during extreme drought.

In Florida, it thrives in conditions that kill most perennials.

Beach Morning Glory: Fast-spreading coastal groundcover

Beach morning glory is native to Florida beaches. It produces edible leaves and spreads faster than sea purslane, making it better for erosion control.

Ipomoea pes-caprae grows as a sprawling vine that spreads across sand. It produces purple flowers and thick, succulent leaves. The young leaves are edible cooked.

I planted beach morning glories along my property line in an area that gets salt spray. Within two years, it'd spread across the entire section.

It stabilizes the sand and requires zero maintenance.

Role: Best for fast-spreading coastal ground cover. Choose beach morning glory over sea purslane if erosion control matters more than constant edible production.

How you'll use it: Young leaves cooked like spinach.

How to grow it:

Plant beach morning glory in full sun. It tolerates pure sand and salt spray. Water occasionally until established, then it survives on rainfall.

How to harvest: Pick young leaves and cook like spinach.

Why it works:

Beach morning glories grow on Florida beaches. It spreads aggressively and stabilizes sand.

If you have coastal property with erosion problems, this plant solves them while producing edible leaves.

Saltbush: The Flood-and-Drought Survivor

Saltbush thrives in areas that flood after heavy rain, then dry out completely for weeks.

Atriplex species produce edible leaves that taste slightly salty. The plant grows as a shrub and handles salt buildup in soil, flooding, drought, and neglect.

My wife and I planted saltbush in a low spot that floods after heavy rain, then dries out completely for weeks. Most plants died there.

Saltbush thrived.

It handles the flooding, survives the drought, and produces edible leaves year-round.

Role: Best for inland areas with salt buildup in the soil and unpredictable moisture. Choose saltbush over sea purslane if you're inland and dealing with flooding followed by drought.

How you'll use it: Cooked like spinach, raw in small amounts for a salty flavor.

How to grow it:

Plant saltbush in full sun. It tolerates wet soil temporarily and thrives in dry soil. Water occasionally until established, then it survives on rainfall.

How to harvest: Pick young leaves anytime. Cook similar to spinach or use raw in small amounts.

Why it works:

Saltbush grows in areas with salt buildup in the soil and unpredictable moisture. It tolerates conditions that kill most perennials.

If you have inland areas with drainage problems and salt issues, saltbush solves them.

Railroad Vine: Dual-Purpose Coastal Plant

Railroad vine produces edible tubers and stabilizes sand on coastal properties.

Ipomoea pes-caprae produces sprawling vines with thick leaves and pink flowers. The tubers are edible cooked and taste similar to sweet potatoes.

Railroad vine handles salt spray, drought, flooding, and pure sand.

Role: Best for coastal sand stabilization with edible tubers. Choose railroad vine over beach morning glory if you want tubers and ground cover.

How you'll use it: Tubers cooked like sweet potatoes.

How to grow it:

Plant railroad vine in full sun. It tolerates pure sand and salt spray. Water occasionally until established.

How to harvest: Dig tubers after the plant's been growing for at least a year. Cook similar to sweet potatoes.

Why it works:

Railroad vine grows on Florida beaches. It spreads fast and stabilizes sand while producing edible tubers.

If you have coastal property with shifting sand, this plant thrives and produces food.

Seagrape: The Coastal Fruit Tree

Seagrape produces edible fruit and handles salt spray better than almost any other fruiting plant.

Coccoloba uvifera grows as a large shrub or a small tree. It produces clusters of grape-like fruit that are edible fresh or made into jelly.

The fruit is tart and seedy but flavorful.

I have a seagrape growing near the coast. No one has ever watered it. I have never fertilized the sea grape. It produces fruit every year.

Birds compete with me for the harvest, but there's enough for both.

Role: Best coastal fruit tree.

How you'll use it: Fresh fruit, jelly, juice.

How to grow it:

Plant seagrape in full sun. It tolerates pure sand and salt spray. Water occasionally until established.

How to harvest: Pick fruit in late summer when it turns purple.

Why it works:

Seagrape grew on Florida's beaches. It handles salt at the cellular level and produces fruit in conditions that kill most fruit trees.

Spanish Bayonet: Edible Flowers in Pure Sand

Spanish bayonet produces edible flowers and survives extreme drought in pure sand.

Yucca aloifolia grows as a tall, spiky plant with sharp leaves. The flowers are edible and taste mildly sweet.

Spanish bayonet survives months without water and tolerates salt spray.

Role: Best for edible flowers in extreme drought.

How you'll use it: Flowers fresh in salads or cooked like squash blossoms.

How to grow it:

Plant Spanish bayonet in full sun. Sharp leaves can cause injury—plant away from walkways. Water occasionally until established.

How to harvest: Pick flowers in spring when they bloom.

Why it works:

Spanish bayonet grew in coastal and arid regions. It stores water and survives extreme neglect.

These Plants Remove Failure from the Equation

Every plant in this chapter survives conditions that would kill everything else in this book.

Salt spray. Months without water. Pure sand. Flooding followed by drought.

These aren't plants you baby. These are plants you plant once and forget about.

They grow.

They produce.

They survive.

In my experience, the plants in this chapter are the ones I recommend to people who say they kill everything.

Because you can't kill these. They're too tough.

I've watched prickly pear survive three months without rain. I've seen sea purslane grow in pure beach sand with constant salt spray. I've planted saltbush in spots where everything else died and watched it thrive.

These plants don't need you. They just need the right spot.

Your problem areas are production zones now

Coastal properties with salt spray. Inland areas with water restrictions. Drought-prone high ground. Flood-prone low spots.

These aren't limitations anymore. They're opportunities to grow the toughest, lowest-maintenance perennials in Florida.

But what about food production specifically? What about perennials that focus on feeding you every single week?

The plants in the next chapter are the grocery store perennials. Edible and medicinal perennials that earn their space by reducing grocery bills. Herbs, perennial greens, and food-producing plants that come back year after year.

Skip the next chapter, and you'll keep spending money at the grocery store on herbs and greens you could grow for free. You'll keep buying basil for three dollars a bunch when you could harvest it fresh every single day. You'll keep buying salad greens when you could walk outside and pick them.

Read the next chapter, and you'll cut your produce bill while harvesting fresh food year-round.

This is where perennials stop being a hobby and start being a food system.

Chapter 8 The Grocery Store Perennials

EDIBLE AND MEDICINAL PERENNIALS THAT EARN THEIR SPACE. HERBS, PERENNIAL GREENS, AND FOOD PRODUCING PLANTS THAT COME BACK YEAR AFTER YEAR AND REDUCE GROCERY BILLS.

E very time you buy fresh herbs, you pay premium prices for plants that die in your fridge.

This chapter is about the perennials that end that habit. You plant them once, then harvest for years. Herbs for dinner, greens for lunch, and medicinal staples you'd normally pay for in a bottle.

In my backyard, we stopped buying fresh herbs years ago. We cook, walk outside, grab what we need, and the plants refill themselves.

Planting more herbs in a rollable raised bed close to our
back door for quick access when cooking.

That's what grocery store perennials mean. They earn their space.

Now, the trap.

Most people try to force sweet basil to behave like a perennial in Florida. It won't.

Florida heat pushes it to flower, seed, and quit. Treat basil as seasonal, then lean on real perennials like Cuban oregano, rosemary, and Tulsi to carry your garden year-round.

The Math That Matters

Fresh herbs at the grocery store cost three to five dollars per bunch. Buy herbs twice a month and you spend seventy to one hundred twenty dollars per year on something that wilts in your fridge.

Perennial herbs cost five to ten dollars per plant. Once established, they produce for years.

Fresh salad greens cost four to six dollars per container. Buy greens once a week and you spend two hundred to three hundred dollars per year.

Perennial greens cost five to ten dollars per plant, or you can grow from cuttings for free. Once established, they produce daily.

Turmeric and ginger cost eight to twelve dollars per pound at the store. One rhizome planted in your yard produces pounds of fresh spice every year.

Turmeric, which I planted and then forgot about. Turns out it doesn't need much attention, just water, and it does its thing.

The plants in this chapter pay for themselves in the first year. After that, everything you harvest is pure savings.

Annual Savings Per Plant

One Cuban oregano plant replaces **sixty dollars per year** in dried herbs

One lemongrass clump replaces **forty dollars per year** in store-bought stalks

One katuk plant replaces **one hundred fifty dollars per year** in salad greens

One moringa tree replaces **one hundred dollars per year** in multivitamins and leafy greens

One turmeric rhizome replaces **eighty dollars per year** in fresh turmeric root

One ginger rhizome replaces **seventy dollars per year** in fresh ginger root

If you only plant three grocery store perennials

If you have time, money, or space for only three edible perennials that'll cut your grocery bill, here's how to choose:

For fresh herbs year-round: Cuban oregano, Mexican tarragon, and rosemary. All three produce constantly and replace expensive store-bought herbs.

For fresh greens daily: Katuk, longevity spinach, and Okinawa spinach. All three produce tender leaves year-round.

For high-value spices: turmeric, ginger, and lemongrass. All three replace expensive store-bought spices.

For maximum savings: moringa, katuk, and Cuban oregano. These three produce more edible biomass per square foot than any other perennials in this chapter.

Now let's get to the plants.

Cuban Oregano: The Herb That Replaces Three Store-Bought Herbs

Cuban oregano isn't actually oregano, but it tastes like a combination of oregano, thyme, and sage. It's one of the most versatile culinary herbs in Florida.

Plectranthus amboinicus produces thick, fuzzy leaves with a strong, savory flavor. It grows as a sprawling groundcover or a bushy plant, depending on how you manage it.

It handles full sun, partial shade, drought, and neglect.

In my backyard, Cuban oregano grows in three different spots. One in full sun. The other under a live oak in partial shade. One in a container.

All three produce constantly. I harvest leaves every time I cook. The plant regrows in days.

My wife uses Cuban oregano in almost every dish that calls for oregano, thyme, or sage. We stopped buying dried herbs three years ago.

This one plant replaced all three.

Role: Best culinary herb for Florida. Replaces oregano, thyme, and sage with one plant.

How you'll use it: Beans, chicken, marinades, roasted vegetables, pizza, pasta sauces.

How to grow it:

Plant Cuban oregano from cuttings in full sun to partial shade. It roots in days. Water occasionally until established. Once established, it handles drought well but grows faster with consistent moisture.

Fertilize lightly every few months.

How to harvest: Cut stems, leaving at least two sets of leaves on the plant. Don't strip it bare. The plant regrows from the remaining leaves.

Use fresh or dried.

Why it works:

Cuban oregano is native to tropical regions with heat and humidity. It thrives in Florida conditions and produces year-round.

Mexican Tarragon: The Licorice-Flavored Perennial Herb

Mexican tarragon tastes like French tarragon but grows as a perennial in Florida heat.

French tarragon doesn't survive Florida summers. Mexican tarragon thrives in them.

Tagetes lucida produces narrow leaves with a distinct licorice flavor. It grows as a bushy plant.

I planted Mexican tarragon in full sun four years ago. I have never replaced it. It produces year-round. I harvest it constantly for chicken dishes, fish, and salad dressings.

Role: Best tarragon substitute for Florida. Use in any recipe that calls for French tarragon.

How you'll use it: Vinegar, salad dressings, fish, chicken, béarnaise sauce.

How to grow it:

Plant Mexican tarragon in full sun with decent drainage. Water regularly until established. Once established, it handles dry spells but grows better with consistent moisture.

Fertilize lightly every few months.

How to harvest: Cut sprigs from the top third of the plant. Leave the lower two-thirds intact. The plant regrows constantly.

Use fresh or dried.

Why it works:

Mexican tarragon grows in hot, dry climates. It handles Florida heat and produces year-round.

Rosemary: The Only Mediterranean Herb That Works Long-Term

Rosemary is one of the few Mediterranean herbs that survives as a perennial in Florida. But it doesn't handle wet feet.

Plant rosemary in raised beds or areas with excellent drainage.

Rosmarinus officinalis produces needle-like leaves with a strong, pine-like flavor.

I killed three rosemary plants before I figured out the drainage issue. I planted them on flat ground. They sat in water after the heavy rain. Root rot killed them.

Then I planted one in a raised bed with fast-draining soil. That plant is still alive five years later.

My wife and I use rosemary constantly. Roasted potatoes. Grilled chicken. Bread.

This one plant produces more than we can use.

Role: Best Mediterranean herb that actually works in Florida. Essential for anyone who cooks.

How you'll use it: Roasted anything, grilled meats, focaccia, roasted potatoes, lamb.

How to grow it:

Plant rosemary in full sun with excellent drainage. Raised beds work best. Water occasionally until established. Once established, rosemary is drought-tolerant.

Don't over-water. Rosemary dies from wet feet faster than drought.

Fertilize lightly once or twice a year.

How to harvest: Cut sprigs from the tips. Don't cut into old, woody growth. The plant regrows from green stems.

Why it works:

Rosemary grows in Mediterranean climates with dry summers and well-draining soil. It tolerates Florida heat if you give it the drainage it needs.

Tulsi Basil: The Perennial Basil That Survives Summer

Tulsi basil, also called holy basil, is the only basil that grows as a perennial in Florida.

Ocimum tenuiflorum tastes different from sweet basil. It has a spicy, clove-like flavor. It's used medicinally in Ayurvedic traditions and works well in teas, soups, and stir-fries.

Tulsi basil doesn't bolt in summer. It keeps producing year-round. It reseeds itself constantly, so even if the original plant dies, new plants come up on their own.

I have tulsi basil growing in three spots in my backyard. I didn't plant the second and third spots. The plant reseeded itself.

Now I have tulsi everywhere. I harvest it constantly for tea.

Role: Best perennial basil for Florida. Use for tea, soups, and stir-fries.

How you'll use it: Tea, broth, soups, stir-fries, Thai curries.

How to grow it:

Plant tulsi basil from seeds or transplants in full sun to partial shade. Water regularly until established. Once established, it handles dry spells well.

Fertilize lightly every few months.

How to harvest: Pinch the top two to three inches of each stem. This encourages branching and more production. Let some flowers go to seed so that the plant reseeds itself.

Why it works:

Tulsi basil is native to tropical India. It grows great in heat and humidity and produces year-round in Florida.

Lemongrass: The Citrus-Flavored Perennial Herb

Lemongrass

Lemongrass produces stalks with a strong lemon flavor. It's essential in Southeast Asian cooking.

Cymbopogon citratus grows as a large clump of grass-like leaves.

I planted one small lemongrass clump four years ago. It's now four feet wide and three feet tall. I harvest stalks constantly for soups, curries, and tea.

The plant keeps producing.

On GrowFitFL, people ask if lemongrass really grows that fast in Florida.

Yes. It does.

One plant becomes a massive clump within two years.

Role: Best citrus-flavored herb for cooking and tea.

How you'll use it: Tea, curry paste, Tom Yum soup, marinades, infused water.

How to grow it:

Plant lemongrass in full sun with decent drainage. Water regularly until established. Once established, it handles dry spells but grows faster with consistent moisture.

Fertilize lightly every few months.

How to harvest: Cut stalks at ground level. Peel the outer layers and use the tender inner stalk. The plant produces new stalks constantly.

Why it works:

Lemongrass is native to tropical Asia. It thrives in Florida's heat and humidity.

Society Garlic: The Garlic-Flavored Perennial

Society garlic produces leaves and flowers with a mild garlic flavor. It replaces garlic in most cooked dishes, but it doesn't produce bulbs.

Tulbaghia violacea produces clumps of narrow leaves and purple flowers. Both the leaves and the flowers are edible.

I planted society garlic in partial shade five years ago. It's spread into a large clump. I harvest leaves anytime I need garlic flavor.

The plant regrows in days.

Role: Best garlic flavor substitute that grows year-round. Doesn't replace garlic bulbs, but works in cooked dishes.

How you'll use it: Stir-fries, soups, pasta, roasted vegetables, compound butter.

How to grow it:

Plant society garlic in full sun to partial shade. Water occasionally until established. Once established, it handles drought well.

Don't fertilize. Society garlic grows in poor soil.

How to harvest: Cut leaves at the base. Use fresh. The flowers are also edible and milder than the leaves.

Why it works:

Society garlic is native to South Africa. It handles heat, drought, and poor soil and produces year-round.

Chives and Garlic Chives: The Perennial Onion Substitute

Chives and garlic chives produce hollow leaves with an onion or garlic flavor. They grow as perennial clumps.

Allium schoenoprasum (regular chives) has a mild onion flavor. Allium tuberosum (garlic chives) has a mild garlic flavor.

Both produce edible flowers.

My wife and I planted both types five years ago. We harvest from them constantly. They regrow within days.

Role: Best perennial onion and garlic substitute for fresh flavor.

How you'll use it: Eggs, baked potatoes, cream cheese, salads, garnish for soups.

How to grow it:

Plant chives in full sun to partial shade. Water regularly until established. Once established, they handle dry spells well.

Fertilize lightly every few months.

How to harvest: Cut leaves two inches above the base. Don't pull them. The plant regrows from the base.

Why it works:

Chives are adaptable to many climates, including Florida. They produce year-round in South and Central Florida.

Turmeric: The High-Value Spice Rhizome

Turmeric produces fresh rhizomes that are far superior to dried turmeric powder. Fresh turmeric has a bright, citrusy flavor that dried powder lacks.

Curcuma longa grows as a leafy plant. The rhizomes develop underground, and farmers harvest them after eight to ten months.

I planted turmeric rhizomes three years ago. Every fall, I harvest pounds of fresh turmeric. We use it in curries, smoothies, and golden milk.

We stopped buying turmeric powder and fresh turmeric root.

Role: Best high-value spice for fresh use. Replaces expensive store-bought turmeric root and powder.

How you'll use it: Curries, smoothies, golden milk, stir-fries, rice dishes.

How to grow it:

Plant turmeric rhizomes in spring in partial shade. Turmeric prefers shade in Florida. Water regularly. The plant needs consistent moisture.

Fertilize lightly every few months.

How to harvest: Dig up rhizomes eight to ten months after planting when the leaves start to yellow and die back. Save some rhizomes to replant.

Fresh turmeric stains everything. Wear gloves.

Why it works:

Turmeric is native to tropical Asia. It thrives in Florida's heat and humidity and produces heavily in partial shade.

Ginger: The Essential Kitchen Spice

Ginger produces fresh rhizomes with a spicy, warming flavor. Fresh ginger is essential in Asian cooking and far superior to dried ginger powder.

Zingiber officinale grows as a leafy plant. The rhizomes develop underground, and harvesters collect them after eight to ten months.

I planted ginger rhizomes in partial shade three years ago. Every fall, I harvest pounds of fresh ginger. We use it constantly.

Stir-fries. Tea. Marinades.

We haven't bought ginger in three years.

Role: Best high-value spice for fresh use. Replaces expensive store-bought ginger root.

How you'll use it: Stir-fries, tea, marinades, ginger ale, baked goods, curries.

How to grow it:

Plant ginger rhizomes in spring in partial shade. Ginger prefers shade in Florida. Water regularly. The plant needs consistent moisture.

Fertilize lightly every few months.

How to harvest: Dig up rhizomes eight to ten months after planting when the leaves yellow. You can also harvest small amounts earlier by carefully digging around the edges without disturbing the main plant. Save some rhizomes to replant.

Why it works:

Ginger is native to tropical Asia. It thrives in Florida's heat and humidity and produces heavily in partial shade.

Moringa: The Nutrient-Dense Multivitamin Tree

Moringa appeared in Chapter 4, but it deserves mention here because it's extremely nutrient dense.

Studies show moringa leaves are a meaningful source of vitamins A, C, calcium, iron, and protein. People have valued moringa as a nutritional supplement for centuries.

In my backyard, I harvest moringa leaves every week. My wife adds them to smoothies, soups, and stir-fries. We use moringa as part of our nutrition strategy.

Nutrition note: Moringa's nutrient density varies based on fresh versus dried leaves and growing conditions. Fresh leaves are milder and work well in cooking. Dried leaves are more concentrated and work well in smoothies.

Role: Best perennial for nutrient density.

How you'll use it: Smoothies, soups, stir-fries, tea.

How to grow it:

See Chapter 4 for full growing instructions.

How to harvest: Pick entire leaflet stems from the top of branches. The more you harvest, the more the plant produces.

Why it works:

Moringa grows faster than almost any other perennial and produces year-round.

Katuk: The Protein-Rich Perennial Green

Katuk appeared in Chapter 5, but it deserves mention here because it's unusually high in protein for a leafy green.

Katuk leaves contain more protein than most greens and have a mild, pleasant flavor that works raw or cooked.

My wife and I harvest katuk leaves daily. We add them to salads, smoothies, and stir-fries.

This one plant produces more greens than we can eat.

Role: Best perennial green for protein content. Replaces expensive salad greens.

How you'll use it: Salads, smoothies, stir-fries, wraps, spring rolls.

How to grow it:

See Chapter 5 for full growing instructions.

How to harvest: Pick leaves from the top six inches of each branch. Don't strip the entire plant. Leave lower leaves intact. The plant regrows constantly.

Why it works:

Katuk produces tender, protein-rich leaves year-round in sun or shade.

What These Plants Have in Common

Every plant in this chapter produces food you'd otherwise buy at the grocery store.

Herbs. Greens. Spices.

They come back year after year. They produce constantly.

In my experience, these are the plants that actually change how you eat. Not because they're exotic. Because they're useful.

Every week, you get a harvest from them. You use what you harvest. You stop spending money at the store.

One of my kids asked me last year why we don't buy herbs anymore. I told him to walk outside and look at the Cuban oregano covering half the side bed.

He got it.

Your grocery bill just got smaller

Walk outside today. Picture yourself harvesting fresh herbs every time you cook. Imagine picking fresh greens for salads every single day. Imagine never buying basil, turmeric, ginger, or salad greens again.

That's what these plants do. They become part of how you eat.

But growing food is only part of what a perennial garden can do. The real transformation happens when you stop thinking about individual plants and start thinking about systems.

The next chapter is about planting techniques. How to dig a hole. Where should you place the plant? How deep to go. What happens in the first thirty days?

Skip it and you'll kill plants that should've thrived. Read it and you'll never lose a plant to bad planting techniques again.

Chapter 9 What Actually Grows Well Together (And Why That Matters More Than You Think)

I spent two hundred dollars learning that companion planting isn't about making your garden look pretty.

It's about making your life easier.

Year one, I planted moringa trees in neat rows. Just moringa. No overstory or ground cover with it. The soil baked hard between waterings. Weeds exploded. I spent hours pulling weeds around trees that should've been taking care of themselves.

Then I watched a video about food forests and guilds (planting communities that support each other), got excited, and crammed every "companion plant" combination I could find into my yard.

Tomatoes with basil. Corn with beans and squash. Marigolds everywhere because the internet said they repel pests.

Half of it died. The other half competed for the same water, the same nutrients, same space.

I wasted money on plants that fought each other instead of helping each other.

That's when I started paying attention to what actually worked in *Florida*.

Not in a permaculture manual written for Oregon. Not in companion planting charts designed for raised bed gardens in California.

Here. In west-central Florida. Where it's ninety-five degrees in July and everything either drowns or dries out.

This chapter is about plant combinations that actually make sense in Florida heat. The ones that reduce work. Save water. Fix nitrogen. Suppress weeds. Confuse pests just enough so that you're not spraying every week.

I'm not giving you theory. I'm giving you what's still alive in my yard after four years of trial and error and realizing that most companion planting advice doesn't translate to subtropical chaos.

The Three Rules That Actually Matter

Most companion planting advice focuses on pest control or nutrient sharing.

That stuff helps, but in Florida, the real question is simpler: does this combination make your life easier or harder?

Here's what I've learned after killing a lot of plants and keeping the ones that actually worked together:

Rule 1: Stack heights, not competition

Plant tall, medium, and ground cover together. Moringa at eight feet. Sweet potato vines at ground level. Something mid-height in between.

This isn't about aesthetics. It's about using all the vertical space so you're growing three layers of food in the same footprint instead of one.

When I planted moringa alone, I got moringa. When I planted moringa with pigeon peas underneath and sweet potato vines as ground cover, I got moringa, beans, and sweet potatoes from the same four-by-four space.

Same water. Same fertilizer. Triple the output.

Rule 2: Nitrogen-fixers earn their spot by feeding everyone else

Legumes (pigeon peas, beans, sunn hemp) pull nitrogen from the air and store it in root nodules. When those roots die back or you chop-and-drop the plants, that nitrogen feeds everything nearby.

I don't fertilize my moringa trees anymore. The pigeon peas planted around them do it for me.

Pigeon peas also produce edible beans, so they're feeding the soil *and* feeding us. That's a plant earning its space twice.

Rule 3: Ground covers suppress weeds, **or you pull weeds forever**

Bare soil in Florida becomes a weed factory within days. Plant a ground cover that spreads fast, and you'll stop spending your weekends pulling crabgrass and dollarweed.

Sweet potato vines, Okinawa spinach, peanut plants — anything that covers ground and out-competes weeds saves you hours of labor.

I planted sweet potato vines under my moringa trees three years ago. I haven't pulled weeds in that area since.

The sweet potatoes are growing. My moringa continues to grow and the weeds lose.

The Moringa Guild (My Most-Used Combination)

This is the combination I've planted more than any other because it works and keeps working.

Tall layer: Moringa tree

Fast-growing. Produces edible leaves year-round. Needs regular harvesting to stay productive.

Mid-layer: **Pigeon peas**

Nitrogen-fixer. Produces edible beans. Grows three to six feet tall. Tolerates partial shade from the moringa canopy.

Ground layer: Sweet potato vines

Edible tubers and greens. Covers bare soil. Suppresses weeds. Handles full sun and partial shade.

I planted this combination in four different spots in my backyard. All four are still productive three years later with minimal input.

Here's what actually happens:

The moringa grows fast. I harvest leaves every week for smoothies and cooking. The more I harvest, the more it branches and produces.

The pigeon peas grow underneath. They fix nitrogen. Their roots feed the moringa. I harvest beans every few months. When pigeon pea branches get woody, I cut them back and let them decompose in place (chop-and-drop). More nitrogen for the moringa.

The sweet potato vines spread across the ground. They block weeds. I harvest sweet potatoes once a year and greens whenever I need them.

I water this combination once or twice a week in the dry season. I don't fertilize. The pigeon peas handle that.

Total cost to set this up:

One moringa tree: $10

Three pigeon pea plants: $15 (or free from seed)

Sweet potato slips: Free (I propagated from grocery store sweet potatoes)

Total cost: $25 or less per guild.

Annual output:

Moringa leaves: Twenty pounds or more per year

Pigeon pea beans: Five to ten pounds per year

Sweet potatoes: Ten to twenty pounds per year

Sweet potato greens: Continuous harvest

This one combination feeds us year-round and pays for itself in two months.

What I Tried That Didn't Work (And Why)

Combination: Moringa + Cassava

Cost: $25 (one moringa, three cassava cuttings)

Problem: Both are tall. Both want full sun. Cassava shaded out the moringa. Moringa grew leggy, trying to compete for light.

Lesson: Don't plant two tall crops together unless you want them fighting.

Combination: Turmeric + Ginger + Galangal (all rhizomes together)

Cost: $40 (bought rhizomes of all three)

Problem: All three need the same thing consistent moisture, partial shade, similar nutrients. They competed for water. Yields dropped across the board.

Lesson: Just because plants have similar needs doesn't mean they should grow together. They might compete instead of complementing.

Combination: Katuk + Longevity Spinach + Okinawa Spinach (all greens together)

Cost: $30 (bought three plants of each)

All three are leafy greens. All three wanted the same light, the same water, the same space. They didn't layer. They crowded each other.

Lesson: Companion planting works when plants use different niches. Three ground covers competing for the same niche is just crowding. Katuk + one of them can work, not both.

The Cassava Hedge (Windbreak + Food + Nitrogen-Fixer)

Cassava grows tall and woody. I planted it along the back fence as a windbreak and privacy screen.

Then I planted pigeon peas in front of the cassava.

Tall layer: Cassava

Windbreak. Privacy. Edible tubers after eight to twelve months.

Mid-layer: **Pigeon peas**

Nitrogen-fixer. Produces beans. Tolerates partial shade from cassava.

Ground layer: Peanuts (raw, store-bought **peanuts)**

Nitrogen-fixer. Ground cover. Suppresses weeds. Handles shade.

This combination does four things:

Blocks wind (cassava)

Produces food (cassava tubers, pigeon pea beans, groundnuts cover)

Fixes nitrogen (pigeon pea, perennial peanut)

Suppresses weeds (perennial peanut)

I planted this along my back fence three years ago. It's still there. Still productive. I harvest cassava tubers once a year, pigeon pea beans every few months, and I haven't pulled weeds along that fence line in two years.

Cost: $20 (cassava cuttings, pigeon pea seeds, perennial peanut plugs)
Labor after establishment: Maybe an hour per year

The Shade Guild (For Under Your Oak Tree)

Most of Florida has a massive oak tree somewhere in the yard. You can't cut it down (it's beautiful and provides shade). You can't grow tomatoes under it (not enough sun).

So you plant shade-tolerant perennials that actually like filtered light.

Canopy: Existing oak tree

You didn't plant it. It's already there. Use it.

Mid layer: Katuk

Shade-tolerant. Produces protein-rich greens year-round.

Ground layer: Okinawa spinach

Shade-tolerant. Spreads as ground cover. Produces edible leaves constantly.

Accent: Turmeric and ginger

Both prefer partial shade. Both produce high-value rhizomes.

I have this combination under the oak tree in my backyard. It's the shadiest spot in my yard and one of the most productive.

The katuk grows three to five feet tall. I harvest leaves every week.

The Okinawa spinach spreads across the ground. I harvest leaves for salads and stir-fries.

The turmeric and ginger grow in clumps between the katuk plants. I harvest rhizomes every fall.

This area produces more food per square foot than most of my full-sun beds. The Florida heat didn't stress the plants. They're thriving in shade.

I water this combination twice a week during the dry season. I fertilize lightly every few months with compost. That's it.

Cost: $35 (katuk plant, Okinawa spinach cuttings, turmeric rhizomes, ginger rhizomes)

Annual output: More greens and spices than I can use

The Front Yard Compromise (Productive + Acceptable to Neighbors)

Most HOAs don't want food forests in the front yards. But they'll tolerate "landscaping."

So you plant productive perennials that look ornamental enough to avoid complaints.

Tall layer: Lemongrass

Ornamental grass appearance. Produces edible stalks. Looks intentional.

Mid-layer: Society garlic

Purple flowers. Looks decorative. Produces edible leaves and flowers with a garlic flavor.

Ground layer: Sweet potato vines (ornamental varieties)
Ornamental foliage. Covers ground. Produces edible tubers (yes, even the ornamental ones).

This combination looks like landscaping. Your neighbors see ornamental grass, flowering plants, and ground cover.

You see lemongrass for tea and curry, garlic-flavored leaves for cooking, and sweet potatoes at the end of the season.

I planted this in the front yard four years ago. Nobody complained. I've harvested from it constantly.

Cost: $25 (one lemongrass clump, three society garlic plants, sweet potato slips)

Neighbor complaints: Zero

Food harvested: Constant

The Pest Confusion Strategy (Does It Actually Work?)

Companion planting folklore says certain plants repel pests or confuse them.

Marigolds repel nematodes. Basil repels aphids. Garlic repels everything.

I've tried most of these combinations. Here s what I've observed in Florida:

What didn't work:
Marigolds didn't stop nematodes from attacking my tomatoes
Basil didn't stop aphids from covering my peppers
Garlic repelled nothing

What might've worked (but I'm not sure):
Planting Cuban oregano near brassicas. I had fewer cabbage worms. Could be the oregano. Could be luck. I kept doing it anyway.

Inter-planting lemongrass with other crops. Anecdotally, areas with lemongrass seemed to have fewer mosquitoes. Again, could be coincidence.

What definitely worked:

Diversity itself. When I planted twenty different crops in the same area instead of rows of one crop, pest pressure dropped. Not because any specific plant repelled pests, but because pests couldn't find their preferred host as easily. They'd land on moringa when they wanted cassava. Land on katuk when they wanted sweet potato.

That's not companion planting folklore. That's just basic ecology. Monocultures get hammered. Diversity spreads the damage.

I don't plant marigolds to repel pests anymore. I plant diversity because it works.

Combinations I'm Still Testing

I don't have answers for everything. Some combinations are still experimental in my yard.

Elderberry + Comfrey

Elderberry grows tall. Comfrey is a deep-rooted nutrient accumulator that supposedly mines minerals and makes them available to nearby plants. I planted this combination two years ago. The elderberry is growing well. The comfrey keeps dying back in the summer heat. I'm not sure yet if this works long-term in Florida.

Papaya + Sunn Hemp

Papaya grows fast but depletes soil quickly. Sunn hemp fixes nitrogen and grows even faster. I planted sunn hemp around young papaya trees last year. The papaya trees look healthier than in previous years. Could be the nitrogen. Could be better rainfall. I'm continuing the experiment.

Muscadine Grapes + Pigeon Peas

Muscadine vines grow on a trellis. Pigeon peas grow underneath as nitrogen-fixers. This combination makes sense on paper. The pigeon peas are thriving, but the muscadines are slow to establish. Not sure yet if this is a winning combination or just two plants tolerating each other.

How to Build Your Own Combinations

You don't need to copy my guilds exactly. You need to understand the principles and adapt them to what you're growing.

Here's how I design plant combinations now:

Step 1: Pick your anchor plant

This is usually a tall perennial that produces something you actually want. Moringa. Cassava. Papaya. Elderberry. Something that'll be there for years.

Step 2: Add a nitrogen-fixer

Pigeon peas. Sunn hemp. Perennial peanut. Beans. Something that feeds the soil so you don't have to fertilize as much.

Step 3: Add a ground cover

Sweet potato vines. Okinawa spinach. Perennial peanut. Anything that suppresses weeds so you're not pulling them every week.

Step 4: Test and adjust

Plant it. Watch what happens. Remove one plant if they compete. If something dies, replace it with something else. If it works, repeat it in other areas.

I've killed dozens of plants figuring out what works. That's normal.

You're not designing a guild once and walking away. You're observing, adjusting, and learning what thrives in your specific yard.

What Companion Planting Actually Gives You

Companion planting in Florida isn't about planting basil next to tomatoes and hoping for magic.

It's about:

Using vertical space, you can grow more food in less area

Planting nitrogen-fixers so you fertilize less

Covering bare soil, so you'll pull fewer weeds—thank me later.

Adding diversity so pest pressure spreads out instead of concentrating

When I planted moringa alone, I got moringa.

When I planted moringa with pigeon peas and sweet potatoes, I got moringa, beans, sweet potatoes, sweet potato greens, nitrogen-fixing roots feeding the soil, ground cover blocking weeds, and a system that mostly takes care of itself.

Same space. Same water. Way less work.

That's what good companion planting does. It makes your garden easier to manage and more productive at the same time.

The combinations I've shared in this chapter work in my yard. They'll probably work in yours if you're anywhere in Florida with similar heat and humidity.

Start with one guild. The moringa combination or the shade guild or the front yard compromise.

Plant it. Watch it. Adjust as needed.

Then plant another one.

Before long, you'll have plant combinations that feed each other, suppress weeds, and produce food year-round with minimal input from you.

And you'll wonder why you ever planted things in rows.

Your Next Step

Walk outside and look at your yard. Find one spot where you could plant a combination instead of a single plant.

It could be located under your oak tree. Maybe it's along a fence line. Maybe it's in a corner that's always full of weeds.

Pick your anchor plant. Add a nitrogen-fixer. Add ground cover.

Plant it. Water it. Let it grow.

Then come back in six months and see what happened.

If it worked, repeat it somewhere else. If it didn't, adjust and try again.

Anyone can practice companion planting without difficulty. It's just paying attention to what works and doing more of it.

Now let's talk about how to actually get these plants in the ground without killing them in the first thirty days.

The next chapter is about planting technique how deep to dig, where to place the plant, what happens when you screw it up, and what to do when you realize the hole you just dug is in the wrong spot.

Part 3: How to Make Them Thrive Long Term

Muscadine grapes - one of my favorite Florida perennials. They handle our heat and humidity better than any other grape, and the kids devour them fresh off the vine. They come back every year and will do so without my help for decades to come.

Chapter 10 The Florida Planting Hole

I n Florida, planting techniques matter more than plant choice.

You can pick the perfect plant. You can place it in a spot with just enough sun and a little afternoon shade. You can water and fertilize it correctly.

And it'll still die if you plant it wrong.

Florida soil isn't soil. It's sand or limestone that drains too fast or not at all. Most planting advice assumes you have real soil with structure, drainage, and buffering capacity.

You don't.

And books that work everywhere else fail you here by omission. They don't lie. They just leave out the Florida-specific details that determine whether plants establish or fail.

The cost of planting wrong isn't just one dead plant. It's the plant itself, the time spent planting it, the water and fertilizer used trying to save it, and the replacement plant you buy next season.

Multiply that by ten plants and you've spent hundreds of dollars learning what this chapter teaches in twenty minutes.

This chapter is about the Florida planting hole. Depth matters. The test for drainage that shows whether the spot will be suitable. The spacing that inhibits fungi. And the first thirty days determine whether the plant establishes or fails.

Get this right, and plants will thrive. Get it wrong and you keep replacing them. I want to help you save some money!

Why Depth Matters More Than You Think

Most gardening advice tells you to plant at the same depth the plant was growing in the container. Dig a hole as deep as the root ball. Set the plant in. Backfill.

Done.

That works in real soil. It fails in Florida sand.

Florida sand drains so fast that water moves through it before roots can access it. If you plant at the same depth as the container, the root ball sits in sand that dries out faster than the plant can drink.

The solution isn't to dig deeper. The solution is to plant higher.

M-4 mango planted on a slight mound, a couple of inches higher than the surrounding soil. Critical for drainage in Florida, standing water will kill these trees fast.

In my backyard, I plant almost everything slightly above grade. The top of the root ball sits one to two inches above the surrounding soil. Then I mulch heavily around it.

This does two things. It improves drainage around the root zone. And it keeps the crown of the plant above the wet zone during heavy rains.

I learned this after killing three katuk plants by planting them at grade. They sat in water after every storm. Root rot killed them.

Then I planted one higher, with the root ball slightly elevated. That plant is still alive five years later.

The Rule: Plant one to two inches above grade in sand. Mound soil or compost around the root ball to create a raised area. Then mulch.

The Visual Checkpoint: If the root ball disappears into the ground, you planted too low.

The Exception: If you're planting in a raised bed with excellent drainage, you can plant at grade. The raised bed already solves the drainage problem.

The Sand Drainage Test: Do This Before You Plant Anything

Before you dig a hole, you need to know if water drains or if it sits.

This takes five minutes. It saves you from planting in spots where plants drown.

Do not skip this.

I do this every time. I don't plant without this test.

Here's the test.

Dig a hole one foot deep and one foot wide. Fill it with water. Wait for it to drain completely. Fill it again.

Time how long it takes to drain the second time.

If it drains in less than one hour: Drainage is excellent. Plant anything.

If it drains in one to three hours, the drainage is acceptable. Most plants will work. Avoid plants that hate wet feet, like rosemary.

If it drains in over three hours: Drainage is poor. Don't plant here without fixing it first. Either build a raised bed, mound soil to create elevation, or choose plants that tolerate wet soil, like elderberry or pickerelweed.

If water sits overnight, this spot floods. Don't plant edibles here unless they specifically tolerate standing water. Plant elderberry, pickerelweed, or sisso spinach. Everything else will drown.

I do this test in every new planting area. It tells me what'll work and what'll fail before I spend money on plants.

In one low spot, water sat for two days. I planted elderberries there. It thrived.

If I'd planted moringa or katuk, they would've died in weeks.

This test is non-negotiable. It's the gatekeeper between success and failure.

Spacing for Airflow: The Fungus Prevention Rule

Florida humidity creates perfect conditions for fungal diseases. Black spot. Powdery mildew. Root rot. Leaf blight.

Fungi thrive in still air and crowded plants. When leaves touch, air doesn't move. Moisture stays trapped.

Fungus blooms overnight.

The solution is spacing.

Most gardening advice tells you to space plants based on their mature size. That works for aesthetics. It fails for disease prevention in Florida.

In Florida, you need more space than the plant tag says. Not because the plants need room to grow. Because they need airflow to stay healthy.

The Rule: Add 50% to the recommended spacing on the plant tag.

If the tag says space plants three feet apart, plant them four to five feet apart.

If the tag says space plants one foot apart, plant them eighteen inches apart.

This feels like wasted space at first. The plants look small and far apart.

But within a year, they fill in. And they stay healthy because air moves through them.

I planted Cuban oregano three feet apart the first time. The plants grew together within months. Leaves touched. Air stopped moving.

Powdery mildew showed up.

I had to prune aggressively to open up the airflow.

The second time, I planted them five feet apart. They still filled in, but air moved between them.

No powdery mildew. No fungal problems.

Spacing prevents disease better than any spray.

And crowding plants doesn't just cause disease. It guarantees replacement costs.

Managing Fruit Trees in Central Florida's Winter

If you're in Central Florida like me, you ll quickly learn that with fruit trees in your food forest, having smaller, more manageable trees planted closer together to help protect each other and create a microclimate during the week or two of winter we receive can be the difference between having mangos in the summer or watching your mango tree start over from cold damage every year.

When reading this, please understand the difference between your perennial vegetables and fruit trees.

Most Central Florida gardeners learn after their first 32-degree night that having their fruit trees closer together here isn't a bad thing. You harvest a greater variety of fruit and increase the likelihood of making it through a couple of sub-freezing hours with less damage.

An outstanding book for anyone to read on fruit trees and spacing, even though it's not a Florida garden book, is *Grow a Little Fruit Tree* by Ann Ralph.

The First 30 Days: What Success Looks Like

The first month after planting determines whether the plant establishes or fails.

Most people think establishment means the plant grows fast.

It doesn't.

Establishment means the roots spread into the surrounding soil and the plant stops relying on the original root ball for water and nutrients.

You can't see this happening. But you can see the signs that it's working.

Week 1: The plant looks the same as it did in the container. Leaves might droop slightly from transplant shock. This is normal. Water every other day. Don't fertilize.

Week 2: Leaves perk up. The plant adjusts to its new environment. Water every other day. Still no fertilizer.

Week 3: New growth appears. This is the first sign that roots are spreading. You can reduce watering to twice a week. Still no fertilizer.

If there's no fresh growth by week three, something's wrong. Ensure the drainage is clear. Check for root rot. Check for pests. Adjust watering.

Don't wait and hope. The plant is telling you it's not establishing.

Week 4: New growth speeds up. The plant has now established roots. You can switch to weekly watering and apply light fertilizer if needed.

What failure looks like: Leaves stay droopy. No fresh growth appears. Leaves yellow or drop. The plant is not establishing itself. Ensure drainage is functioning. Check for root rot. Check for pests. Adjust watering.

In my experience, most plant failures happen in the first thirty days because people either over-water, underwater, or fertilize too early.

I killed moringa cuttings by fertilizing them in week one. I was trying to help them grow faster. The fertilizer burned the roots before they'd had time to spread.

The plants died.

Now I wait thirty days before I add any fertilizer. The plants establish better. Growth is slower at first, but it's sustainable.

How to dig a hole effectively.

Here's the step-by-step process I use for every plant.

Step 1: Do the drainage test.

Dig a test hole. Fill it with water. Time for the drainage. Decide if the spot works.

Don't skip this.

Step 2: Dig the planting hole.

Getting ready to plant a 15-gallon mango tree. The hole looks deep, but I'm going to backfill most of this dirt to raise the tree 2-3 inches above ground level, critical for drainage in Florida.

Dig a hole twice as wide as the root ball, but only as deep as the root ball. You want width for the roots to spread. You don't want depth that traps water.

Step 3: Rough up the sides.

If you're planting in sand, this doesn't matter much. If you're planting in soil with clay or hardpan, scratch the sides of the hole with a shovel. This prevents the hole from becoming a bowl that traps water.

Step 4: Set the plant higher.

Place the root ball in the hole so the top sits one to two inches above the surrounding soil. Backfill around the sides. Don't bury the crown.

If the root ball disappears into the ground, you planted too low.

Step 5: Create a watering basin.

Mound soil or compost in a ring around the plant, about six inches from the stem. This creates a shallow basin that holds water and directs it toward the roots instead of running off.

Step 6: Mulch heavily.

Add three to four inches of mulch around the plant. Keep mulch two inches away from the stem to prevent rot. Mulch moderates soil temperature, holds moisture, and suppresses weeds.

Step 7: Water deeply.

Saturate the soil with water. This settles the soil around the roots and eliminates air pockets.

That's it.

Simple. Effective. And completely different from the advice that tells you to dig deep and plant at grade.

What happens if you skip these steps?

If you plant at grade in sand, water drains away before roots can access it. The plant dries out faster. You water more. The plant still struggles because the root zone can't hold moisture.

If you plant in a spot with poor drainage without testing first, the plant sits in water after every rain. Root rot sets in. The plant died slowly.

You think it's a disease or a pest. It's not. It's drainage.

If you crowd plants together, fungus moves in. You are spraying. You trim. You are in a continuous struggle against it.

Because you created the conditions that fungi love.

And you guarantee replacement costs.

If you fertilize in the first thirty days, you burn roots that haven't spread yet. The plant dies or becomes stunted.

You blame the plant. It wasn't the plant. It was the timing.

I've made every one of these mistakes. I've killed plants by planting them wrong. I've lost money, time, and confidence because I followed generic advice instead of adjusting for Florida conditions.

The Florida planting hole does not involve complications. But it's specific.

And it's the difference between plants that establish and plants that fail.

This chapter alone saves you hundreds of dollars

Every plant you lose costs money. The plant you purchased from the nursery. The time spent planting and caring for the plants. The replacement plant you buy next season.

If this chapter prevents you from losing three plants, it's saved you more than the cost of this book.

In my backyard, I stopped losing plants once I figured out the planting hole. I stopped digging deep. I started planting high. I tested the drainage before planting. I spaced for airflow. I waited thirty days before fertilizing.

The plants established faster. They grew stronger. They required less maintenance.

Not because they were better plants. Because I planted them correctly.

My wife noticed before I did. She walked through the beds one afternoon and said, "Nothing's dying anymore."

She was right.

The Hole Is Only Half the Battle

Planting is something you know how to do. You understand how to space. You can test the drainage. The first thirty days should be clear to you.

But planting at the right depth in the right spot doesn't matter if you plant at the wrong time.

Timing in Florida is backward. The seasons you think are planting seasons aren't. The months you think are growing months are survival months.

And if you plant in May thinking it's spring, you're setting plants up to fail before they even establish.

The next chapter is about timing. When to plant in Florida and when not to. It explains why people often overrate spring planting and how fall and winter planting quietly outperform spring planting.

Get the timing right and plants establish fast. Get it wrong and they struggle for months before dying in the summer heat.

Chapter 11 Timing Is Everything

WHEN TO PLANT IN FLORIDA AND WHEN NOT TO. WHY SPRING PLANTING IS OFTEN OVERRATED AND HOW FALL AND WINTER QUIETLY OUTPERFORM IT.

You can dig the perfect hole.

You can plant at the right depth.

You can space for airflow and test drainage before you plant.

And the plant will still die if you plant it at the wrong time.

Timing in Florida is backward. Spring isn't the beginning of the growing season. Summer isn't when plants thrive. Fall isn't when things slow down. Winter isn't dormancy.

In my backyard, I killed plants for two years because I planted them in spring. The plants struggled through the summer heat. The plants showed stress. They attracted pests. They died before the fall arrived.

Then I started planting in the fall and winter. The same plants thrived.

Not because they were different plants. Because I planted them when Florida was ready for them.

If you've been planting in spring and wondering why nothing works, therefore.

Why Spring Planting Fails More Often Than It Succeeds

Spring in Florida isn't the beginning of the growing season. It's the countdown to survival mode.

Temperatures climb into the 90s. Humidity spikes. Afternoon storms dump rain daily.

And you're asking a plant that just went into the ground to establish roots while fighting extreme heat, high humidity, and fluctuating moisture.

Plants don't thrive under those conditions. They survive.

And survival isn't the same as establishment.

When you plant in spring, you have six to eight weeks before the summer heat arrives. If the plant doesn't establish strong roots in that window, it goes into summer stressed.

Stressed plants attract pests. They develop fungal diseases. They stop growing. Many of them die.

Spring planting works only if everything goes right.

Florida rarely cooperates.

I planted moringa in April, thinking I was giving it the entire growing season. By June, the plant was struggling. Leaves yellowed. Growth slowed. Aphids moved in.

The plant barely survived the summer. It didn't start thriving until October, when the temperatures dropped.

The following year, I planted moringa in October. By the time summer arrived, the plant had six months of root growth. It handled the summer heat without stress.

It grew through summer instead of surviving it.

Spring planting in Florida isn't impossible. But fall planting works better for most perennials.

Why Fall and Winter Are the Best Planting Seasons

Fall and winter are when Florida wakes up.

Temperatures drop into the 70s and 80s. Humidity eases. The relentless afternoon storm tapers off. Soil stays moist but not saturated.

And plants establish roots without fighting extreme heat.

This is the proper planting season in Florida.

When you plant in the fall, the plant has six months of mild weather to establish before summer heat arrives. Roots spread. The plant builds strength.

By the time summer hits, the plant is ready. It doesn't just survive the summer. It grows through it.

In my backyard, almost everything I plant now goes in the ground between October and February. Moringa, Katuk. Longevity spinach. Cuban oregano.

All of them establish faster, grow stronger, and produce more when planted in fall or winter.

My wife and I planted katuk in November three years ago. By March, the plant was bushy and producing leaves. By summer, the plant had established itself so well that the heat didn't slow it down.

That plant is still producing heavily today.

Compare that to the katuk I planted in April. The summer was a struggle for it. It barely grew. It took a full year to reach the size that the fall-planted katuk reached in six months.

Fall and winter planting gives plants time to establish without stress.

That's the advantage.

The Florida Planting Calendar by Season

This isn't a suggestion. This is the planting rhythm that works in Florida.

Fall: October and November

Best planting season for most perennials.

Plant moringa, katuk, chaya, longevity spinach, Okinawa spinach, Brazilian spinach, Cuban oregano, Mexican tarragon, tulsi basil, lemongrass, and most edible perennials.

Temperatures are mild. Soil stays moist. Plants establish fast. By the time the summer heat arrives, they're ready.

Avoid planting nothing. Fall is the best time to plant almost everything.

Winter: December, January, February

Second-best planting season.

Plant the same perennials in the fall. Winter temperatures in South and Central Florida are mild enough for most tropicals to establish.

In North Florida, wait until late February for frost-sensitive plants like moringa.

Temperatures are cool, but not cold in most of Florida. Plants establish slowly but steadily. By spring, they're ready to grow fast.

Avoid planting: In North Florida, avoid planting frost-sensitive tropicals like moringa, cassava, and papaya until late February. Wait until the last frost has passed.

Spring: March, April, May

Acceptable planting season if you plant early and fast.

Plant in March or early April if you must plant in spring. Give plants six to eight weeks to establish before the summer heat arrives.

Spring works better for some perennials than for others. Prickly pear cactus, agave, and other succulents handle spring planting well because they tolerate heat stress.

Moringa, katuk, and leafy greens struggle more because they need time to establish before the heat hits.

If you plant in spring, water more frequently. Mulch heavily. Watch out for pests and fungal diseases.

And accept that establishment will take longer than it would in fall.

Avoid planting in late May. By late May, summer heat is here. Plants struggle to establish themselves . Wait until fall.

Summer: June, July, August, September

Don't plant.

Summer is survival mode for plants. Extreme heat, high humidity, daily rain, and pest pressure create the worst conditions for establishment.

Plants you put in the ground in summer spend all their energy trying to survive. Roots don't spread. Growth slows or stops. Pests and fungal diseases move in.

Most plants that go into the ground in summer die before fall.

I've tried planting in the summer. It doesn't work.

Even tough plants like moringa and longevity spinach struggle when planted in June or July.

By fall, they look worse than they did when I planted them.

The only exception is plants that are already established in containers. If you have a moringa or katuk growing in a pot and you want to transplant it to the ground, summer transplanting can work if you water heavily and provide shade for the first few weeks.

But even then, fall transplanting works better.

Avoid planting: Everything. Wait until October.

Planting Windows by Plant Type

Not all perennials follow the same rules. Here's the breakdown.

Tropicals (Moringa, Katuk, Chaya, Cassava, Longevity Spinach, Okinawa Spinach)

Best planting time: October through February
Acceptable planting time: March through early April
Avoid: May through September

Tropicals establish best in fall and winter when temperatures are mild. They can handle spring planting if you plant early, but they struggle with summer heat during establishment.

Perennial Herbs (Cuban Oregano, Mexican Tarragon, Rosemary, Tulsi Basil, Lemongrass, Society Garlic, Chives)

Best planting time: October through February
Acceptable planting time: March through April
Avoid: May through September

Herbs establish fast in fall and winter. They can handle spring planting better than tropicals, but fall planting still gives better results.

Rhizomes (Turmeric, Ginger)

Best planting time: February through April
Avoid: October through January, June through September

Turmeric and ginger are the exceptions. They need warm soil to break dormancy. Plant them in late winter or early spring when soil temperatures rise.

They grow through spring and summer, then go dormant in fall. Harvest in the fall when the leaves die back.

Succulents (Prickly Pear Cactus, Agave, Spanish Bayonet)

Best planting time: March through May
Acceptable planting time: October through February
Avoid: June through September

Succulents tolerate heat stress better than other perennials. They establish well in spring because they don't need consistent moisture.

Fall planting works too, but spring gives them the entire growing season to spread roots.

Natives (Coontie, Saw Palmetto, Elderberry, Yaupon Holly, Passion-flower, Persimmon)

Best planting time: October through February

Acceptable planting time: March through April

Avoid: May through September

Natives follow the same pattern as tropicals. Fall and winter planting gives them time to establish before the summer heat.

What happens if you plant at the wrong time?

If you plant in late spring or summer, the plant goes into the ground during the most stressful time of year.

Heat stresses the plant. Humidity invites fungus. Daily rain creates standing water. Pests explode in population.

The plant can't establish roots fast enough to handle the stress. Growth slows. Leaves yellow. Pests move in. Fungal diseases develop.

You water more.

You fertilize more.

You spray for pests.

Nothing helps because the problem isn't care. The problem is the timing.

Timing determines whether a plant establishes fast or struggles for months. And struggling plants rarely catch up.

They stay stunted. They stay stressed. And many of them die.

How to know if you've missed the window

If you're reading this in May and thinking you missed the fall planting window, you did.

But that doesn't mean you can't act.

Here's what to do if you're outside the ideal planting window.

If it's late spring (May): Wait until October. Don't plant now. Use the time to prepare planting areas, improve soil, and plan your layout.

Waiting isn't inaction. It's strategic timing.

If it's summer (June through September): Wait until October. Use these months to research plants, source materials, and get ready.

You're not wasting time. You're setting yourself up for success in the fall.

If it's early spring (March or early April): Plant now if you must. Water frequently. Mulch heavily. Accept that establishment will take longer.

If it's fall or winter (October through February): Plant everything. This is your window.

In my backyard, I do most of my planting in October and November. Those two months account for 80% of the perennials I put in the ground. The other 20% go in between December and February.

I almost never plant between May and September anymore. The results aren't worth the effort.

Waiting until fall isn't laziness. It's discipline. It's choosing success over impatience.

Timing Matters More Than Most People Think

You can choose the right plant.

You can plant it in the right spot.

You can dig the perfect hole with excellent drainage.

And it'll still struggle if you plant it at the wrong time.

Timing in Florida isn't about following a calendar. It's about understanding when plants can establish without fighting extreme conditions.

And in Florida, that means fall and winter for most perennials.

The plants I put in the ground in October thrive. The plants I put in the ground in May are struggling.

Same plants. Same care. Different timing. Completely different results.

My wife figured this out before I did. She stopped planting anything between May and September two years before I did. When I asked why, she said, "Because they all die."

She was right.

But planting at the right time is only part of the System

You know when to plant. You know when not to plant. You know which plants have flexible windows and which plants need specific timing.

But even perfectly timed plants struggle if you don't know how to manage them through Florida's summer.

Because summer in Florida isn't a growing season. It's a survival test.

The next chapter is about the summer survival guide. Managing the rainy season, heat stress, fungi, pests, and root rot. How to keep plants alive when Florida feels hostile.

Skip it, and your fall-planted perennials will struggle through their first summer.

Read it and they'll grow through summer instead of just surviving it.

Chapter 12 The Summer Survival Guide

MANAGING THE RAINY SEASON, HEAT STRESS, FUNGUS, PESTS, AND ROOT ROT. HOW TO KEEP PLANTS ALIVE WHEN FLORIDA FEELS HOSTILE.

S ummer in Florida isn't a growing season. It's a test.

Extreme heat. Relentless humidity. Daily afternoon storms that dump a lot of rain fast. Standing water. Fungal explosions. Pest populations that double overnight.

And plants that were thriving in spring suddenly struggled to survive.

Florida summers are brutal. And if you don't know how to manage it, you lose plants.

In my backyard, I lost more plants in summer than in any other season during my first three years of gardening. Not because the plants were weak. Because I didn't understand what summer required.

I watered when I should've let them dry out. I fertilized when I should've stopped. I ignored the early signs of stress until it was too late.

Then I learned how summer works. I learned when to intervene and when to leave plants alone. I learned how to read stress before it became fatal.

And I stopped losing plants.

This chapter is about summer survival. Tips for keeping plants safe during the rainy season. You should know how to recognize heat stress and how to address it. How to prevent fungus before it starts. How to handle pests without spraying constantly. And how to spot, root rot before it kills.

Get this right and your plants will grow through the summer. Get it wrong and they die in the season they should be thriving.

The rainy season changes everything

Summer in Florida is the rainy season. June through September. Afternoon storms almost every day. Then sun. Then another storm the next day.

The problem isn't the rain. The problem happens after the rain.

If your planting area has poor drainage, water sits around the roots and root rot sets in. If your planting area has excellent drainage, the top layer dries out between storms and roots near the surface stress.

You can't control the rain. But you can control how your planting area handles it.

The rule: Check moisture before you water. Push your finger two inches into the soil. If it's wet, don't water it. If it's dry, water lightly.

In summer, most plants need less help than you think, and too much help kills them.

I used to water every morning during summer, thinking I was helping. I was drowning the plants.

Afternoon storms had already saturated the soil. I was adding more water to soil that had no room for it.

Root rot killed three moringa plants before I figured this out.

Now I only water in summer if we go three days without rain and the soil is dry two inches down. Most of the time, I don't water at all.

The plants do better without my help.

Summer Survival Checklist

Use this as your daily reference during the summer months:

Check soil moisture before watering (finger test two inches down, every time)

Water early morning only (never at night; leaves need time to dry)

No heavy fertilizer June through September (light feeding only in June, then stop)

Prune for airflow (remove lower leaves, thin dense growth, keep air moving)

Remove infected leaves immediately (fungus spreads fast; act the day you see it)

Tolerate small pest levels (intervene only when damage spreads; let predators work)

Plant high, mulch, protect the crown (keep mulch two inches away from stems so the crown stays dry)

Two-Minute Diagnosis for Wilting

When a plant wilts, use this process before you do anything:

Step 1: Check the soil **two inches down**

Wet: Don't water. Assume heat stress or drainage issues.
Dry: Water deeply.

Step 2: Check the timing

Wilts midday only and perks up at night: Heat stress. Leave it alone.
Wilts morning and evening too: Drought stress or root issue.

Step 3: If wet soil plus wilting plus yellowing leaves

Suspect root rot. Stop watering immediately and inspect the roots.

Heat stress looks different from drought stress

Plants wilt in summer. When you see wilting, your first instinct is to water it.

Sometimes that helps. Sometimes it makes things worse.

You need to know the difference between heat stress and drought stress. They look similar, but they require opposite responses.

Heat stress: Leaves droop in the middle of the day when temperatures peak. The soil is moist. The plant perks up in the evening when temperatures drop.

This isn't a drought. This is the plant reducing surface area to conserve water during peak heat.

What to do: Nothing. Don't water. The plant is managing itself. It'll recover when temperatures drop.

Drought stress: Leaves droop in the morning or evening when temperatures are moderate. The soil is dry two inches down. The plant doesn't perk up at night.

This is drought. The plant needs water.

What to do: Water deeply. Saturate the soil around the root zone. Then check soil moisture daily until the plant recovers.

In my backyard, I have moringa that wilts every afternoon in July and August.

Constant watering was my panicked reaction that first summer. The soil remained saturated. The plant developed root rot.

The following summer, I checked the soil, found it moist, and watched the plant perk up every evening. I stopped watering.

The plant thrived.

Learning to read the difference between heat stress and drought stress saves plants.

Watering heat-stressed plants drowns them. Ignoring drought-stressed plants kills them.

Know which one you're looking at.

Fungus Is Inevitable. Losing the plant is optional.

Florida summers create perfect conditions for fungal diseases. High humidity. Warm temperatures. Wet leaves. Still air.

Powdery mildew. Black spot. Leaf blight. Root rot. All of them thrive in the summer.

You can't eliminate fungus. But you can reduce it to manageable levels with three strategies.

Strategy 1: Space plants for airflow.

When plants are too close, they prevent air circulation. Moisture stays trapped on leaves. Fungus blooms overnight.

If you plant with proper spacing, fungal pressure is lower. If you didn't, prune aggressively now. Remove lower leaves. Thin dense growth. Let air move through the plant.

Strategy 2: Water in the morning, never at night.

If you water in the evening, the leaves stay wet all night. That's eight hours of perfect fungal conditions.

If you water in the morning, the leaves dry by afternoon.

In my backyard, I water before 9 AM if I water at all in summer. By noon, the leaves are dry.

Strategy 3: Remove infected leaves immediately.

Fungus spreads. One infected leaf becomes ten in days.

If you see powdery mildew, black spot, or leaf blight, remove the infected leaf immediately. Throw it in the trash, not the compost. Then check the plant daily for new infections.

I used to leave infected leaves thinking the plant would fight it off. The infection spread. By the time I acted, the infection had covered half the plant.

Now I remove infected leaves the day I see them. Fungal pressure remains low.

Pests Explode in Summer. Your response matters more than the pests.

Aphids, whiteflies, spider mites, caterpillars, and beetles all thrive in Florida summers. Pest populations double every few days in warm, humid conditions.

Most people see pests and immediately spray them .

That works short term. It creates bigger problems long term.

Spraying kills pests and beneficial insects. Ladybugs, lacewings, predatory wasps, spiders. These insects eat pests.

When you spray, you kill both populations. The pests come back faster because they reproduce faster. The predators don't.

You spray again. The cycle repeats. You're now spraying every week.

The better strategy is to tolerate low pest levels and let predators do the work.

The Rule: Only intervene when pest damage is severe or spreading fast.

A few aphids on fresh growth? Leave them. Ladybugs will find them.

Aphids covering every leaf and stunting growth? Intervene.

Caterpillars eating a few leaves on passionflower? Leave them. That's the point of passionflower.

Caterpillars stripping an entire plant to bare stems? Intervene.

The first summer I sprayed aphids the moment I saw them. They came back fast, and I ended up spraying weekly.

Next summer I left them alone for a few days. Ladybugs showed up, and the aphids crashed.

That lesson changed how I garden.

Now I only spray if pest damage is severe and spreading fast. Most of the time, I do nothing.

The pest populations stay low because predators keep them in check.

When you need to intervene: Use insecticidal soap or neem oil. Both kill pests on contact but break down quickly. Spray in the evening when beneficial insects are less active, but avoid spraying when plants are heat-stressed. Test a small section first. Focus on the undersides of leaves where pests hide.

Root Rot Is the Silent Summer Killer

Root rot kills more plants in Florida summers than any other problem. And most people don't realize it's happening until the plant is dead.

Root rot doesn't start with visible symptoms on leaves. It starts underground.

Roots sit in saturated soil. Oxygen gets cut off. Roots turn brown and mushy. They stop taking up water and nutrients.

By the time leaves yellow, the root system already sustains damage.

Early signs of root rot:

Leaves yellow, starting from the bottom of the plant

New growth looks stunted or discolored.

The plant wilts even though the soil is wet

The stem feels soft near the soil line

The plant doesn't respond to fertilizer

If you catch it early, stop watering and let the soil dry. Improve drainage by raising the planting area, mounding, or moving the plant. Remove yellow leaves and watch for fresh growth.

If it's advanced: Dig it up and check the roots. Healthy roots are firm and light-colored. Rotten roots are dark and mushy.

If most roots are rotten, discard the plant and replant in a well-draining spot. If only some roots are rotten, trim the damaged roots and replant higher with fast drainage.

I lost two katuk plants to root rot before I learned to recognize the early signs. Both times, I saw yellowing leaves and assumed nutrient deficiency. I fertilized.

The plants got worse.

By the time I dug them up, the roots were completely rotten.

Now I check drainage before I plant. I plant high to improve drainage. And if I see yellowing leaves on a plant in wet soil, I stop watering immediately and inspect the roots.

I haven't lost a plant to root rot in three years.

Root rot is preventable. Plant in areas with good drainage. Plant high. Don't over-water during the rainy season. Check the soil moisture before you water.

What Summer Success Looks Like

Summer in Florida isn't about explosive growth. It's about steady survival that leads to fall growth.

Plants that handle summer well don't grow fast. They flourish. Fresh growth appears every week, but slowly. Leaves stay green. Pest damage is minor. When fungal issues occur, they are isolated and manageable.

Plants that struggle in summer stop growing. Leaves yellow. Pest damage spreads. Fungal infections cover multiple leaves.

The plant looks worse in August than it did in June.

In my backyard, summer is when I focus on maintenance, not growth. I prune infected leaves. I check for pests weekly. I water only when it's necessary. Fertilizing is not something I do.

I let the plants manage themselves over the summer with minimal help from me.

By October, the plants that handled summer well explode with growth. The plants that struggled in summer take months to recover.

Summer is the test. Fall is the reward.

The One Thing You Should Not Do in Summer

Don't fertilize heavily during the summer.

Fertilizer pushes growth. Growth in summer attracts pests. Fast-growing tender leaves are pest magnets. Aphids, whiteflies, and caterpillars target fresh growth first.

Fertilizer also increases water demand. Plants growing fast need more water. In summer, when rain is unpredictable and heat is extreme, that creates stress.

If you must fertilize in summer, use slow-release organic fertilizers in small amounts. Compost. Worm castings. Fish emulsion diluted to half strength.

Apply once in June. Then stop until October.

I used to fertilize moringa every month, thinking I was helping it grow faster. The plants grew faster, then aphids swarmed them. Growth stalled.

Now I fertilize lightly in June and not again until October. The plants grow slower in summer, but they stay healthier.

By fall, they're stronger and ready to grow fast.

Summer isn't the enemy. It's just different.

Summer in Florida feels hostile. The heat is extreme and t humidity is relentless. Florida rain is unpredictable, but the pests are constant. The fungus is inevitable.

But summer isn't the enemy. It's just different.

And once you learn how to manage it, plants survive and even thrive.

The plants in my backyard that struggled in summer during my first few years now grow through summer without major issues.

Not because they're tougher plants. Because I stopped fighting summer and started working with it.

Before watering, I check the soil moisture. I recognize heat stress and leave it alone. I space plants for airflow. I remove infected leaves immediately. I tolerate low pest levels. I don't fertilize heavily. I plant high to prevent root rot.

Those strategies keep plants alive. And plants that stay alive through summer explode with growth in the fall.

One of my kids asked me last summer why I wasn't watering the moringa when it looked droopy. I told him to come back at sunset and check it again.

He did. The plant perked up and was growing.

He got it.

But survival isn't enough

Managing the rainy season is something you know how to do. You are familiar with how to identify heat stress. Preventing fungi and managing pests is something you know how to do. You know how to spot root rot before it kills.

But keeping plants alive isn't the same as keeping them productive.

And productivity requires feeding them correctly.

The next chapter is about fertilizing perennials in Florida. How to feed them without burning roots or triggering disease. Slow, steady, climate-appropriate nutrition that builds long-term fertility instead of creating short-term growth spurts that attract pests.

Skip it and your plants will survive, but underperform.

Read it and they'll produce heavily year after year without constant intervention.

Chapter 13 Feeding Perennials Without Killing Them

FERTILIZING IN FLORIDA WITHOUT BURNING ROOTS OR TRIGGERING DISEASE. SLOW, STEADY, AND CLIMATE APPROPRIATE NUTRITION.

F ertilizer kills more plants in Florida than a lack of fertilizer.

That sounds backwards. Most people think plants need constant feeding. More fertilizer equals more growth. Bigger harvests. Healthier plants.

In Florida, more fertilizer equals burned roots, pest explosions, and fungal diseases.

Florida sand has no buffering capacity. When you add fertilizer to sand, it doesn't get diluted or held in reserve like it does in real soil. It concentrates around the roots.

Too much and the roots burn. The plant dies from overfertilization, not starvation.

Add fertilizer during the wrong season and you push tender growth that attracts every pest in the neighborhood. Add it too often and you create lush, weak growth that invites fungus.

In my backyard, I killed more plants with fertilizer than I ever did with neglect during my first two years. I was following the instructions on the bag.

Feed every month. More for heavy feeders.

I burned roots. I triggered pest infestations. Suddenly, plants that were thriving declined.

Then I learned how fertilizer actually works in Florida sand. Knowing when to feed and when to stop was something I learned. I learned which fertilizers work and which one's burn. I learned to read what plants need instead of following a schedule.

The plants I feed lightly and infrequently produce more than the plants I used to feed heavily every month.

Not because they need less nutrition. Because they can actually use what I give them.

This chapter is about fertilizing perennials in Florida without killing them. When to feed. When to stop. Which fertilizers work? How much is enough? And how to read a plant so you know what it actually needs instead of guessing.

Why Sand Changes Everything About Fertilizer?

Real soil has clay particles, organic matter, and cation exchange capacity. Those things hold onto nutrients and release them slowly.

When you add fertilizer to real soil, it gets buffered. The soil releases it gradually. Plants access it as they need it.

Florida sand has none of that.

Sand is pure quartz. It has no ability to hold nutrients. When you add fertilizer to sand, it either leaches through with the next rain or stays concentrated where you put it.

If it leaches through, you wasted money and the plant never got fed.

If it stays concentrated, the salt concentration around the roots becomes toxic. The roots burn. The plant stops taking up water even though the soil is moist. Leaves yellow. Growth stops.

The plant dies from fertilizer burn, not nutrient deficiency.

Therefore, the fertilizer schedule that works in other states fails in Florida. You're not working with soil that can buffer excess. You're working with sand that amplifies mistakes.

The Rule: In Florida, less is more. Light applications. Slow-release products. Long intervals between feedings.

In my backyard, I fertilize perennials three to four times per year. Not monthly. Not biweekly.

Three to four times. And the plants produce heavily.

When to Fertilize and When to Stop

Fertilizer timing in Florida is as important as the fertilizer itself.

When you feed at the wrong time, you stimulate growth while the plant should conserve energy. Feed during summer and you create tender growth that attracts pests. Feed during establishment, and you'll burn roots before they spread.

The Florida Fertilizer Calendar

October and November: This is the primary feeding window. Temperatures drop. Plants shift into active growth mode. They can actually use the nutrients you give them. Apply slow-release organic fertilizer. Compost. Worm castings. This feeding carries plants through fall and winter.

February and March: Second feeding window. Plants are growing actively. Weather is mild. Apply slow-release organic fertilizer or compost tea. This feeding supports spring growth.

June: Optional light feeding. Only if plants show signs of nutrient deficiency. Use half-strength liquid fertilizer or a light top-dressing of compost. Then stop. Don't feed again until October.

July, August, September: Don't fertilize. Summer is survival mode. Fertilizer pushes growth that attracts pests and increases water demand during the most stressful time of year. Let plants coast through summer on stored nutrients.

During the first 30 days after planting, don't fertilize. Roots need time to spread before you add nutrients. Fertilizing too early burns roots that haven't established yet.

I used to fertilize every month, thinking I was helping. I fertilized in July. The plants pushed tender growth. Aphids swarmed.

I fertilized the newly planted moringa in week one. The roots burned. The plant died.

Now I fertilize in October, February, and sometimes lightly in June.

That's it. Three times per year. The plants produce more than they did when I was fertilizing monthly.

Which fertilizers work in Florida

Not all fertilizers are equal in Florida sand.

Fast-release synthetic fertilizers burn roots in sand. Their concentration happens too fast. They leach through too fast. They create short bursts of growth followed by crashes.

Slow-release organic fertilizers work because they feed soil biology first, then plants. They degrade over time. They don't concentrate to toxic levels. They build long-term fertility rather than creating short-term growth spurts.

Fertilizers that work in Florida:

Kitchen composter that's been a game-changer. We've got four kids producing food scraps and 125– plants needing compost. Overnight processing beats waiting months for a traditional pile to break down.

Compost: The safest fertilizer for Florida. It nourished the soil biology. This enhances structure. It releases nutrients slowly. Apply one to two inches as a top-dressing around plants twice per year.

Worm castings: Nutrient-rich, impossible to overuse, feeds beneficial microbes. Apply as a top dressing or mix into compost tea. Use liberally.

Milorganite: heat-treated biosolids. Slow-release nitrogen, iron, and micronutrients. Apply according to package directions, but reduce the frequency. Once in October, once in February. That's enough.

Fish emulsion: Liquid organic fertilizer. Fast-acting, but not burning if diluted correctly. Use at half strength. Apply once per month during active growth periods only. Stop in summer.

Composted manure: chicken, cow, horse. All work well if fully composed. Apply as a topdressing. Don't use fresh manure. It burns.

Slow-release synthetic fertilizers: Osmocote, Nutricote. These work in Florida if used sparingly. The coating releases nutrients gradually. But they're expensive and still leach faster in Florida heat and rain than in other climates. Use half the recommended rate.

Fertilizers that fail in Florida:

Fast-release synthetic fertilizers: Miracle-Gro, any water-soluble chemical fertilizer. These burn roots in sand. They leak through in days. They create growth spurts that attract pests. Avoid them.

High-nitrogen fertilizers during summer: Even organic ones. Nitrogen pushes tender growth. Tender growth attracts aphids, whiteflies, and caterpillars. Use balanced fertilizers or low-nitrogen products.

In my backyard, I use compost and worm castings as my primary fertilizer. I top-dress with compost in October and February. I add worm castings to planting holes. I use fish emulsion diluted to half strength once per month in fall and spring.

I don't use synthetic fertilizers anymore. The plants are healthier, and I spend less money.

How to Know What Your Plants Actually Need

Most people fertilize on a schedule without checking if plants actually need it.

Plants tell you what they need if you know how to read them.

Signs of nutrient deficiency:

Yellowing leaves starting from the bottom of the plant: Nitrogen deficiency. The plant is pulling nitrogen from old leaves to feed fresh growth. Apply compost or fish emulsion.

Yellowing leaves with green veins: Iron deficiency. Common in alkaline soils or limestone-based areas. Apply chelated iron or sulfur to lower the pH.

Purple tint on leaves or stems: Phosphorus deficiency. Common in acidic soils. Apply bone meal or rock phosphate.

Stunted growth with dark green leaves: Phosphorus lockout, usually from high pH. Lower the pH with sulfur or plant in amended areas.

Overall pale color, slow growth: General nutrient deficiency. Apply balanced organic fertilizer or compost.

Signs you're over-fertilizing:

Leaf tips and edges turn brown and crispy: Fertilizer burn. Salt buildup around the roots. Stop fertilizing. Flush the soil with water. Let the plant recover.

Lush, dark green growth that wilts easily: Too much nitrogen. The plant is growing too fast to support itself. Stop fertilizing. Let the plant harden off.

Rapid growth followed by pest infestations: Too much nitrogen. You're creating pest magnets. Stop fertilizing. Let predators catch up.

In my backyard, I check the plants every week. The color of the leaf is what I'm looking at. I look at the growth rate. I look for signs of deficiency or excess.

Then I adjust feeding based on what I see, not what the calendar says.

Most of the time, plants don't need fertilizer. They need better soil, better drainage, or better watering.

Fertilizer doesn't fix those problems. It makes them worse.

The Finger Test for Fertilizer Needs

Here's the simplest way to know if your plants need fertilizer.

Look at the plant. Is it growing? Is it producing? Are the leaves green? Is new growth appearing regularly?

If yes, don't fertilize. The plant is getting what it needs.

If no, check these things first before you add fertilizer:

Is the soil moist but not saturated? Over-watering and under-watering both cause symptoms that look like nutrient deficiency.

Is drainage good? Poor drainage causes yellowing leaves that look like nitrogen deficiency but are actually root stress.

Are there pests? Aphids and whiteflies cause yellowing and stunted growth. Fertilizing makes pest problems worse, not better.

Is it summer? Plants slow down in summer. Slow growth in July is normal. It's not a nutrient deficiency.

Has the plant been in the ground for less than six months? New plants focus on root growth, not top growth. They look slow. That's establishment, not starvation.

If you have checked all of those and the plant is still showing clear nutrient deficiency symptoms, then fertilize lightly.

I used to fertilize every plant on the same schedule. Some needed it. Most didn't.

Now I fertilize based on what the plant shows me. Three times per year is when some plants get fed. Some get fed once. Some don't get fed at all because the compost mulch is enough.

How Much Is Enough

In Florida, half the recommended rate is usually enough.

Fertilizer labels assume real soil with buffering capacity. Florida sand has no buffering. The recommended rate often burns roots or leaches through without being used.

The Rule: Start with half the recommended rate. Watch the plant. If it responds well, stay there. If it shows continued deficiency, increase slightly. If it shows burning, reduce further.

I follow this rule with every fertilizer. Milorganite says four pounds per 100 square feet. I use two pounds. Fish emulsion says one tablespoon per gallon. I use half a tablespoon.

Compost has no burn risk, so I use it liberally. But anything with concentrated nutrients gets cut in half.

In ten years, I've never had a plant fail from under fertilizing using this rule. I've had multiple plants fail from over-fertilizing when I followed the label rates.

Special Cases: Plants That Need More or Less

Most perennials thrive on light, infrequent feeding. But a few have specific needs.

Heavy feeders that need more: Moringa, katuk, turmeric, ginger, lemongrass. These produce a lot of biomass fast. They benefit from feeding in October, February, and June. Use compost, worm castings, or fish emulsion.

Light feeders that need almost nothing: prickly pear cactus, agave, coontie, saw palmetto, yaupon holly. These evolved in nutrient-poor soils. They don't need fertilizer. Compost mulch once per year is enough. Over-fertilizing creates weak, leggy growth.

Acid-loving plants: blueberries, azaleas, some ferns. These need an acidic pH to access iron and other micronutrients. Use sulfur to lower the pH. Mulch with pine straw. Avoid lime.

Alkaline-tolerant plants: Most perennials in this book tolerate a wide pH range. If you're on limestone-based soil, use chelated iron for plants showing iron chlorosis.

In my backyard, I feed moringa and katuk more than I feed prickly pear or coontie. I adjust based on the plant, not a universal schedule.

What happens if you over-fertilize

If you over-fertilize in Florida, the plant tells you fast.

Leaf tips turn brown and crispy. Growth is lush and dark green, but weak. The plant wilts even though the soil is moist. Pests swarm fresh growth. The plant is declining rapidly.

What to do if you over-fertilized:

Stop fertilizing immediately. Flush the soil with water to leach out excess salts. Water deeply for three to five days in a row, letting water run through the root zone. Then let the soil dry out slightly and resume normal watering.

Remove damaged leaves. They won't recover. Let the plant produce fresh growth.

Don't fertilize again for at least three months. Let the plant recover.

I over-fertilized Cuban oregano with Miracle-Gro when I first started gardening. Leaf tips turned brown. The plant stopped growing.

I flushed the soil. I stopped fertilizing. It took two months for the plant to recover.

I haven't used fast-release synthetic fertilizer since.

Fertilizer Isn't a Cure for Severe Conditions

Fertilizer doesn't fix poor drainage. This does not correct over-watering. It doesn't fix pest infestations. It doesn't fix bad placement.

Most problems that look like nutrient deficiencies are actually problems with water, drainage, pests, or placement.

Before you fertilize, check everything else first.

Does the plant receive adequate sun or shade? Is drainage good? Is watering appropriate? Are pests present?

If those things are wrong, fertilizer makes the problem worse, not better.

In my backyard, I've seen yellowing leaves from over-watering, poor drainage, aphid infestations, and root rot.

Fertilizer fixed none of those problems. Adjusting water, improving drainage, managing pests, and improving planting techniques fixed them.

Fertilizer is the last thing you add, not the first thing you try.

The System That Works

Here's the fertilizing system I use in my backyard.

October: Top-dress all perennials with one to two inches of compost. Add worm castings to heavy feeders like moringa and katuk.

February: Apply fish emulsion diluted to half strength to actively growing plants. Skip slow growers and light feeders.

June: Check plants for nutrient deficiency symptoms. If present, apply light compost or diluted fish emulsion or chicken manure. If not, skip feeding entirely. I will provide some "chop and drop" from either Mexican sunflower, pigeon pea, or banana leaves for mulching and slow feeding.

July through September: No fertilizer. Let plants coast through the summer.

New plantings: No fertilizer for the first 30 days. After 30 days, apply light compost or worm castings.

That's it. Simple. Minimal. Effective.

The plants in my backyard produce heavily. They keep themselves healthy and don't burn. My plants don't attract pest swarms, and they grow steadily without excessive vegetative growth.

I spend less money on fertilizer than I did when I was feeding monthly. The plants perform better.

And I kill nothing with overfertilization anymore.

My wife noticed before I did. She said the plants looked healthier in year three than they did in year one, even though I was doing less.

She was right.

Feeding Isn't the Hard Part

But fertilizing correctly only matters if you're also pruning correctly. Because unpruned perennials become leggy, overgrown, and unproductive. They stop producing leaves in the places you can reach. They waste energy on growth you don't want.

The next chapter is the month-by-month Florida perennial calendar. Exactly what to prune, plant, fertilize, mulch, or leave alone each month. This is the most referenced chapter in the book. The one you'll come back to every month to check what needs doing and what you should ignore.

Skip it and you'll guess your way through the year. Read it and you'll know exactly what to do and when to do it.

Chapter 14 The Month by Month Florida Perennial Calendar

EXACTLY WHAT TO PRUNE, PLANT, FERTILIZE, MULCH, OR LEAVE ALONE EACH MONTH.

E very month, Florida gardeners ask the same question: What should I be doing right now?

Most gardening calendars don't work in Florida. They tell you to prune in early spring. Plant in April. Fertilize in May. Harvest in summer.

Those timelines assume temperate seasons that don't exist here.

Florida operates on a different calendar. What you should do in January isn't what books from other states tell you to do. What you should ignore in July is exactly what those same books tell you is critical.

This calendar assumes Central and South Florida conditions. In North Florida, shift frost-sensitive planting back two to four weeks based on local frost dates.

This chapter is the Florida perennial calendar. Month by month. What to prune. What to plant. What to fertilize. What to mulch. And what to leave alone.

In my backyard, I keep this calendar on my phone. I check it at the beginning of every month. It tells me what actually needs doing and what I can skip.

It prevents me from wasting time on tasks that don't matter in Florida and reminds me of the tasks that do.

This is the chapter you'll reference more than any other. Use it. Follow it. And your perennials will produce more with less effort every year.

But here's what most gardening calendars won't tell you: some months, you won't do anything because you're busy. Kids are home from school. Work is insane. Life happens.

This calendar accounts for that. It tells you what matters and what can wait.

January

Reality Check: Kids are back in school after winter break. You have time. Weather is perfect. This is a wonderful month to actually get things done.

What to Plant: Everything except turmeric and ginger.

January is one of the best planting months in Florida. Some years we have wild cold storms hit that will set this schedule back a few weeks but on average for much of central and all of south Florida temperatures are mild. Soil stays moist. Plants establish fast without fighting heat.

If it is a winter storm, just give it a couple weeks after the last storm has passed.

Plant moringa, katuk, chaya, longevity spinach, Okinawa spinach, Cuban oregano, Mexican tarragon, rosemary, tulsi basil, lemongrass, society garlic, chives.

In North Florida, wait until late February for frost-sensitive tropicals like moringa if frost is still possible.

What to Prune: Prune perennials that grew leggy or overgrown during fall.

Cut back moringa, katuk, and chaya by one-third to one-half to encourage bushy growth. Remove dead or damaged leaves from all perennials. Thin dense growth to improve airflow.

What to Fertilize: Nothing.

Plants are growing steadily on nutrients from October feeding. If you must feed, use light compost tea or diluted fish emulsion on heavy feeders only.

What to Mulch: Refresh mulch around plants that look thin.

Add two to three inches of wood chips, shredded leaves, or compost. Keep mulch two inches away from stems.

What to Leave Alone: Don't over-manage.

January plants are establishing or thriving. Let them do their thing. Check for pests weekly but don't intervene unless damage is severe.

What Actually Gets Done: I plant most of my new perennials in January. I prune moringa hard to keep it bushy. I refresh mulch in areas where it's broken down.

The rest? I leave it alone and let plants grow.

Time required: Three to four hours spread across the month.

Busy parent friendly? Yes. Kids are in school. Weather is comfortable. You can get outside without dying of heat.

February

Reality Check: Still pleasant weather. Still manageable schedule. February is your last chance to get things planted before spring heat arrives.

What to Plant: Continue planting perennials.

February is the last ideal planting month before spring heat arrives. Plant anything you want established before summer.

Start planting turmeric and ginger rhizomes. Soil temperatures are warming. Rhizomes break dormancy faster in late winter than in fall.

What to Prune: Prune spring-flowering perennials after they bloom.

Remove spent flowers and cut back leggy growth. Prune passionflower if it's overgrown its support. Thin Cuban oregano and other herbs to improve airflow.

What to Fertilize: This is your second fertilizing window.

Apply compost, worm castings, or diluted fish emulsion to actively growing perennials. This feeding supports spring growth.

Use half the recommended rate. Skip slow growers and light feeders like prickly pear, agave, and coontie.

What to Mulch: Check mulch levels. Add more if needed.

Mulch prevents weeds, and moderates soil temperature as spring heat arrives.

What to Leave Alone: Don't over-fertilize. One light application is enough.

Don't prune tropicals like moringa too hard this late. Let them grow into spring.

What Actually Gets Done: I fertilize in February. I top-dress with compost and apply diluted fish emulsion to moringa, katuk, and lemongrass. I plant turmeric and ginger. Because the passionflower attempts to overtake the fence, I prune it.

Sometimes I skip fertilizing if I'm swamped. Plants survive fine without it. They just grow a little slower.

Time required: Two to three hours spread across the month.

Busy parent friendly? Yes. Still comfortable weather. Kids in school. Doable.

March

Reality Check: The weather is warming up. Spring break might happen. You might not get much done. That's fine.

What to Plant: Early March is acceptable for planting if you must plant in spring.

After mid-March, wait until October. The window is closing fast. Summer heat is weeks away.

Plant succulents like prickly pear, agave, and Spanish bayonet. These tolerate spring heat better than tropicals.

What to prune: Prune any perennials that need shaping before summer.

Once summer heat hits, pruning stresses plants. Do it now if it needs doing.

What to fertilize: Nothing.

Plants are coasting on February feeding. Don't push growth this close to summer.

What to mulch: Refresh mulch if needed.

Mulch is critical going into summer. It moderates soil temperature and holds moisture.

What to Leave Alone: Avoid planting tropicals after mid-March. Don't fertilize. Don't prune heavily.

Let plants prepare for summer on their own.

What Actually Gets Done: I stop planting in March. I refresh mulch everywhere. I prune anything that looks scraggly before summer locks in.

If spring break happens, and the kids are home, I skip everything except mulch. Mulch matters. The rest can wait.

Time required: One to two hours if you do anything at all.

Busy parent friendly? Barely. Spring break chaos. Get mulch down if you can. Everything else is optional.

April

Reality Check: It's getting hot. You're tired. The yard is transitioning to summer mode whether or not you help it.

What to Plant: The planting window is closed. Wait until October.

What to prune: Light pruning only.

Remove dead or damaged growth. Thin dense areas to improve airflow going into summer. Don't prune heavily. The plant needs its leaves to handle summer heat.

What to fertilize: Nothing.

Fertilizing now pushes tender growth that'll get hammered in summer.

What to mulch: Check mulch levels. Add more if thin.

Mulch is your best tool for helping plants survive the summer.

What to Leave Alone: Avoid planting. Don't fertilize. Don't prune heavily.

April is the preparation month. Get mulch in place, improve airflow, and let plants coast into summer.

What Actually Gets Done: I check mulch. Sometimes I add more. Sometimes I don't.

I prune for airflow if I remember. I watch plants transition from spring growth to summer survival mode.

Honestly? April is when I do less because it's getting too hot to care.

Time required: One hour maximum. Maybe zero.

Busy parent friendly? Yes, because you barely have to do anything.

May

Reality Check: School's ending soon. It's hot. You're not doing yard work in May unless something is actively dying.

What to Plant: The planting window is closed. Wait until October.

What to prune: Remove dead or damaged leaves only.

Don't prune for shape. The plant needs all the healthy leaves it has to handle summer.

What to fertilize: Nothing. Fertilizing now is a mistake.

What to mulch: Mulch is critical now.

Check every plant. Add more mulch if the levels are low. Mulch moderates soil temperature and holds moisture during the hottest months.

What to Leave Alone: Everything.

May is when you stop managing and start observing. Let plants handle summer. Your job is to check soil moisture, watch for severe pest damage, and remove infected leaves.

That's it.

What Actually Gets Done: I shift into a summer mode in May. I check the soil moisture before I water. I remove any leaves showing fungal infections.

The rest? I ignore it. Plants either survive the summer or they don't. I'm not standing outside in ninety-degree heat trying to micromanage perennials.

Time required: Thirty minutes per week checking moisture and removing diseased leaves.

Busy parent friendly? Yes, because you're not doing much. Also, no, because kids are home from school and demanding attention.

June

Reality Check: Summer vacation. Kids are home. It's brutally hot. You're surviving, not thriving. Your plants are too.

What to Plant: Survival mode only. Wait until October.

What to prune: Remove infected leaves immediately.

Fungi spread fast in summer. Don't prune for shape. Only remove diseased or dead growth.

What to Fertilize: Optional light feeding only if plants show clear nutrient deficiency.

Use half-strength fish emulsion or light compost. Then stop. Don't feed again until October.

What to mulch: Check mulch levels.

Mulch breaks down fast in summer heat and moisture. Refresh if needed.

What to Leave Alone: Don't prune heavily. Don't fertilize unless a deficiency is obvious.

Let plants coast through the summer. Check the soil moisture before you water. Most plants need less help than you think.

What Actually Gets Done: I check soil moisture maybe twice a week. I remove fungal leaves when I see them. I tolerate pests unless damage is spreading.

I don't fertilize. I do not prune. I plant nothing.

June is survival mode for me and the plants.

Time required: Twenty minutes per week. Sometimes zero if it rains.

Busy parent friendly? No, kids are home. It's hot. Nobody wants to be outside. You do the bare minimum and call it good.

July

Reality Check: Peak summer. Peak chaos. Four kids are home all day. Ninety-five degrees by 10am. You're not gardening. You're watering if you remember.

What to Plant: Survival mode only. Wait until October.

What to prune: Remove infected leaves only.

Don't prune for shape or size. The plant needs its leaves.

What to fertilize: Nothing. July is peak stress. Fertilizing now creates problems.

What to mulch: Check mulch. Add more if thin.

What to Leave Alone: Everything.

July is the hottest, most humid month. Plants are in survival mode. Your job is to check soil moisture, remove infected leaves, and watch for severe pest damage.

That's it. Don't over-manage.

What Actually Gets Done: Almost nothing.

I check plants weekly when I'm outside with the kids. I remove infected leaves if I see them. Most of the time I don't even water because afternoon thunderstorms handle it.

July is the month I do the least in my entire gardening year.

Time required: Fifteen minutes per week. Maybe.

Busy parent friendly? Absolutely not. But that's fine because the plants don't need you, anyway.

August

Reality Check: Still summer. Still chaos. You're counting down until school starts again.

What to Plant: Survival mode only. Wait until October.

What to prune: Remove infected leaves only.

What to fertilize: Nothing.

What to mulch: Check the mulch. Add more if needed.

What to Leave Alone: Everything. August is still in survival mode. Don't push plants. Let them coast.

What Actually Gets Done: Same as July. I check the plants when I remember to. I removed the infected leaves. I let the afternoon rain do most of the watering.

I'm not accomplishing anything significant in August. I'm waiting for fall.

Time required: Fifteen minutes per week.

Busy parent friendly? Still no. Still hot. Still chaotic.

September

Reality Check: School is starting. Hallelujah. You have time again. The weather is still brutal, but you can see the light at the end of the tunnel.

What to Plant: The planting window opens in October. Wait two more weeks.

What to prune: Light pruning to prepare for fall growth. I typically prefer doing most of the pruning after our last frost possibilities. I like my trees to have a little extra on them going into January so by February, I can start thinking about pruning.

You have to live in North or Central Florida to fully understand that logic. All it takes is a couple of frosty nights to wipe out years of growth that could have been better protected if you had not pruned and allowed each tree to help the other with wind protection.

Remove dead leaves and spent growth. Thin, dense areas to improve airflow. Don't prune heavily yet. Wait until October.

What to fertilize: Nothing. Wait until October.

What to mulch: Check the mulch. Add more if needed.

Mulching in the fall helps plants transition from summer survival to fall growth.

What to Leave Alone: Avoid planting. Don't fertilize.

September is a transition month. Let plants shift from survival mode to growth mode on their own.

What Actually Gets Done: I plan fall planting in September. I order seeds and cuttings. I prepare planting areas. I do light pruning to clean up plants before fall growth starts.

But I don't plant or fertilize yet. I wait until October.

September is planning month, not action month.

Time required: One to two hours preparing for October.

Busy parent friendly? Yes. Kids are back in school. The weather is slightly less brutal. You can think about gardening again without wanting to die.

October

Reality Check: This is it. Your main event. October is when everything happens. Weather is perfect. You have time. Plants are ready to grow.

What to Plant: Everything.

October is the best planting month in Florida. Temperatures drop. Soil stays moist. Plants establish fast. This is your primary planting window.

Plant moringa, katuk, chaya, longevity spinach, Okinawa spinach, Brazilian spinach, Cuban oregano, Mexican tarragon, tulsi basil, lemongrass, society garlic, chives, and all other perennials except turmeric and ginger.

What to prune: Prune heavily now in South Florida, Central Floridaprun but be cautious of January freezes. North Florida, I highly suggest waiting until after your last frost to prune. This isn't garden textbook advice; this is one Central Florida backyard grower to another who understands what 3-5 days of cold can do to your investment.

Cut back moringa, katuk, and chaya by one-third to one-half. Remove leggy growth. Thin dense plants. Shape everything. Pruning now triggers bushy fall and winter growth.

What to Fertilize: This is your primary fertilizing window.

Top-dress all perennials with one to two inches of compost. Add worm castings to heavy feeders like moringa and katuk. This feeding carries plants through fall and winter.

What to mulch: Refresh mulch everywhere.

Add three to four inches around all plants. Keep mulch two inches away from the stems.

What to Leave Alone: Nothing. October is action month.

Plant. Prune. Fertilize. Mulch. This is when you set up for success for the next six months.

What Actually Gets Done: Everything.

October is my busiest month. I plant eighty percent of my new perennials. I prune everything heavily. I top-dress with compost. I refresh mulch.

By the end of October, the garden is reset and ready for fall and winter growth.

I spend more time in the yard in October than I do in June, July, and August combined.

Time required: Eight to twelve hours spread across the month. More if you're planting a lot.

Busy parent friendly? Yes. Weather is perfect. Kids are in school. You can actually work outside without suffering. Take advantage of this month.

November

Reality Check: Still pleasant weather. Still manageable. November is your second chance to get things done before the holidays hit.

What to Plant: Continue planting.

November is the second-best planting month. Temperatures are perfect. Plants establish fast.

What to prune: Prune anything you missed in October. Be mindful of the winter to come when pruning.

After November, let plants grow through winter without heavy pruning.

What to Fertilize: If you missed the October feeding, apply compost now.

Otherwise, skip it. One fall feeding is enough.

What to mulch: Check mulch levels. Add more if needed.

What to Leave Alone: Don't over-fertilize. But get some compost down or whatever you choose to use to feed your plants this month.

One application in October or November is enough for the next few months.

What Actually Gets Done: I finish fall planting in November. I planted anything I couldn't get in the ground in October. I check mulch everywhere. I do light pruning if needed.

Otherwise, I let the plants grow.

November is cleanup month. Finishing what October started.

Time required: Three to four hours spread across the month.

Busy parent friendly? Yes, until Thanksgiving week. Then all bets are off.

December

Reality Check: Holidays. Family visiting. Kids are home from school for two weeks. You might get some light harvesting done. That's about it.

What to Plant: Continue planting if needed.

December is still a good planting month in South and Central Florida. Avoid planting frost-sensitive tropicals like moringa if North Florida expects frost.

What to prune: Light pruning only. North and Central Florida check weather reports; you might not want to prune and allow your trees to hold the extra weight going into a cold stretch.

Remove dead or damaged growth. Let plants grow through the winter.

What to fertilize: Nothing. Plants are coasting on October feeding.

What to mulch: Check mulch levels. Add more if thin.

What to Leave Alone: Don't over-manage.

December plants are thriving. Let them do their thing. Harvest as needed.

What Actually Gets Done: I coast in December.

Plants are growing well. I harvest moringa leaves, katuk, Cuban oregano, and longevity spinach constantly. I don't fertilize or prune. I let plants grow and produce.

December is harvest month, not a work month.

Time required: Thirty minutes per week harvesting. Maybe an hour if you plant something.

Busy parent friendly? Depends on your family's holiday chaos. For me, it's harvest only. No major projects.

The Florida Perennial Year at a Glance

October through February: Build the system

Plant. Prune. Fertilize. Mulch. This is when you set up for success for the year.

March through April: Prepare for Stress

Finish planting early. Stop fertilizing. Refresh mulch. Let plants shift into summer survival mode.

May through September: Observe, Don't Interfere

Check soil moisture. Remove infected leaves. Tolerate pests. Let plants survive summer with minimal intervention.

Best planting months: October, November, December, January, February

Worst planting months: May, June, July, August

Best fertilizing months: October, February

Worst fertilizing months: May, June, July, August, September

The Busy Parent Reality Check

Here's what actually happens when you have four kids, a job, and a perennial garden:

October: You get stuff done. Weather is perfect. Kids are in school. You plant, prune, fertilize, and mulch as if your life depends on it.

November: You finish what October started. You're still motivated. The weather is still good.

December: Holidays hit. You harvest. That's about it.

January: Kids are back in school. You plant some more. You prune. You feel productive again.

February: You fertilize if you remember. You plant turmeric and ginger. You're still engaged.

March: You add mulch because you know summer is coming. Everything else is optional.

April: You check the mulch. Maybe you prune. Maybe you don't.

May through August: Survival mode. You water if it doesn't rain. Removing diseased leaves is what you do. You ignore pests unless they're destroying everything. You spend as little time outside as possible because it's hot and the kids are home.

September: You plan for October. You get excited about fall planting. You prepare beds and order plants.

October: The cycle starts again.

This is the real calendar. Not the idealized version. The one that accounts for school schedules, work deadlines, and that nobody wants to be outside in July.

What Actually Matters vs. What You Can Skip

Here's what I've learned after five years of following this calendar with four kids and a full-time job:

What matters:

Planting in October and November. This sets up the entire year. Miss this window and you're planting in summer (a bad idea) or waiting another year.

Mulching before summer. Do this in March or April, and your plants will survive summer with way less water.

Removing diseased leaves in summer. Fungus spreads fast. Pull infected leaves immediately.

Fertilizing in October. One good feeding in the fall carries plants through winter and spring.

What you can skip without consequences:

February fertilizing. Nice to have, not critical. Plants grow fine without it.

March pruning. Helpful for shaping, but unnecessary. Plants survive summer either way.

May through August, anything except watering and disease removal. Seriously. Do the minimum. Plants handle summer on their own.

December, anything except harvesting. Let plants grow. Enjoy the production.

The pattern you should see

If you follow this calendar, you'll notice a pattern.

Fall and winter are action months. Plant. Prune. Fertilize. Mulch. This is when you set up the system.

Spring is transitioning. Finish planting early. Prepare for summer. Stop fertilizing. Refresh mulch. Let plants shift into summer survival mode.

Summer is an observation. Refrain from planting. Don't fertilize. Don't prune heavily. Check soil moisture. Remove infected leaves. Tolerate pests. Let plants survive summer with minimal intervention.

This pattern repeats every year. And every year, the plants that follow this calendar produce more with less effort.

This Calendar Prevents Mistakes

The biggest mistakes Florida gardeners make happen because they're doing the wrong tasks at the wrong time.

Planting in June. Fertilizing in July. Pruning heavily in August. Following advice from temperate climate books.

This calendar prevents these mistakes. Every month, it informs you of what to do and what to skip. It removes guesswork. It removes wasted effort.

It focuses your energy on the tasks that actually matter in Florida.

In my backyard, I stopped losing plants once I started following this calendar. I stopped planting in the summer. I stopped fertilizing during stress periods. I stopped pruning when plants needed their leaves.

And the plants responded by growing stronger and producing more.

My wife keeps a printed copy of this calendar in the garage. She checks it more than I do now. When I ask if something needs doing, she walks to the garage, checks the month, and comes back with the answer.

Works every time.

The Months You'll Feel Guilty (And Why You Shouldn't)

June, July, and August, you're going to feel you're neglecting your garden.

Watering isn't happening every day. You are not applying fertilizer. You're not pruning. You're barely paying attention.

Everyone else is posting on social media about their beautiful summer gardens. You're inside with the AC on, looking out the window at plants that look stressed and wondering if you're doing this wrong.

You're not.

Florida summers are survival mode for plants and people. The plants that make it through summer are the ones that can handle neglect.

The ones that die? They would not work long-term, anyway.

I killed dozens of plants before I accepted this. I kept trying to "help" them through summer with extra water, extra fertilizer, and extra attention.

All I did was stress them more.

Now I leave them alone in the summer. I check the soil moisture. I remove diseased leaves. That's it.

And the plants that survive produce harder in fall and winter than they ever did when I was micromanaging them.

So when you feel guilty in July about doing nothing, remember: doing nothing is the right call.

The Calendar Adjusts to Your Life

Some years, I plant heavily in October. Other years, I skip October entirely because work is insane, or someone is sick, or life happens.

The calendar doesn't break if you miss a month.

Miss October planting? Plant in November or January instead.

Forget to fertilize in October? Feed in February.

Skip pruning in October? Prune in January.

The calendar is a guide, not a rule. It shows you the ideal windows, but it also shows you the backup windows.

Use it when you can. Skip it when you can. Your garden will survive either way.

The difference between following this calendar perfectly and following it loosely is maybe ten percent more production. The difference between following it loosely and ignoring it completely is plant death and frustration.

Get the big things right, and the small things won't matter as much.

But the calendar is only part of the system

You know what to do each month. You know when to plant, prune, fertilize, and mulch. You know when to leave plants alone.

You know what matters and what you can skip when life gets chaotic.

But following the calendar only works if you're growing the right plants in the first place.

And the right plants for Florida aren't the ones you see in every other gardening book.

The next chapter is about the five plants that prove you've figured Florida out. These aren't beginner plants. These are plants that only thrive once someone truly understands Florida gardening.

Not because they're difficult. Because they require everything you've learned so far to work together.

Skip it, and you'll have a functional garden. Read it and you'll have proof that you've mastered Florida perennial gardening.

Part 4: Confidence and Mastery

Chapter 15 The Five Plants That Prove You've Figured Florida Out

PERENNIALS THAT ONLY THRIVE ONCE SOMEONE TRULY UNDERSTANDS FLORIDA GARDENING.

S ome plants forgive mistakes. They grow despite poor drainage, inconsistent watering, and bad timing.

They're beginner plants. They're what you start with when you don't know what you're doing yet.

The plants in this chapter aren't those plants.

These are the perennials that only thrive when you get everything right. Not because they're difficult. Because they require everything you've learned so far to work together.

Timing. Drainage. Spacing. Pruning. Feeding. Summer management. All of it.

When these plants thrive in your yard, you know you've figured Florida out.

Why These Five?

I chose these five plants deliberately. Each one fails for a different reason.

Together, they test every major Florida gardening skill: drainage, restraint, timing, ecosystem tolerance, and systems thinking.

Cassava tests drainage. Rosemary tests restraint. Turmeric tests timing. Passionflower tests ecosystem thinking. Moringa tests systems-level understanding.

When one or more of these plants thrives in your yard, you have proof that you understand Florida gardening beyond following instructions.

In my backyard, these were the plants that failed when I started gardening and succeeded once I understood Florida's rules. Each one taught me something.

And when they finally thrived, I knew I was no longer fighting Florida. I was working with it.

When your neighbor asks how you got these plants to thrive, you've graduated.

Now you're teaching Florida, not fighting it.

Cassava: The Plant That Tests Your Drainage

Cassava produces starchy tubers that can weigh five to ten pounds each. It grows year-round in South and Central Florida. It handles heat, humidity, and poor soil.

But it punishes poor drainage.

Cassava roots sit underground for eight to twelve months before harvest. If water sits around those roots during that time, they rot.

The plant survives, but the tubers turn to mush. You harvest nothing.

When cassava thrives and produces heavy tubers, it means you planted in a spot with excellent drainage. It means you did the drainage test before you planted. It means you planted high.

In my backyard, I planted cassava in flat ground twice. The plants grew and looked healthy. When I dug up the tubers, they were rotten.

I was planting in spots that flooded after heavy rain.

The third time, I planted cassava in a mounded area with fast drainage. When I harvested eight months later, I pulled five pounds of clean, firm tubers from one plant.

Cassava doesn't tolerate drainage mistakes. When it produces, you know your drainage is right.

Rosemary: The Plant That Tests Your Restraint

Rosemary is one of the few Mediterranean herbs that survives as a perennial in Florida. But it only survives if you leave it alone.

Rosemary needs excellent drainage and infrequent watering. It needs you to resist the urge to help.

Most people kill rosemary by over-watering. They see the plant in Florida heat and assume it needs water. They water daily.

Root rot sets in. The plant dies.

Rosemary thrives on neglect. Water it once at planting. Then leave it alone. Let it survive on rainfall.

Only water if we go two weeks without rain and the leaves curl.

When rosemary thrives in your yard, it means you learned restraint. It means you stopped over-watering. It means you trusted the plant to handle Florida on its own.

In my backyard, I killed three rosemary plants by watering them too much. The fourth rosemary went into a raised bed with excellent drainage.

I watered it once at planting. Then I left it alone.

It's been alive for five years. I water it twice a per year during extended dry spells.

The plant thrives because I learned to stop helping.

Rosemary fails quickly when you try to help. When it survives year after year, you know you've mastered restraint.

Turmeric: The Plant That Tests Your Timing

Turmeric produces fresh rhizomes that are far superior to dried turmeric powder. But it only produces if you plant it at the right time.

Plant turmeric too early, and the rhizomes rot in cold soil before they sprout. Plant turmeric too late, and the plant doesn't have enough time to develop large rhizomes before it goes dormant in the fall.

The window is narrow. Late February through April.

Soil temperatures need to be warm enough to trigger growth but early enough that the plant has six to eight months to develop rhizomes.

When turmeric thrives and produces heavy rhizomes, it means you understood Florida's backward calendar. It means you planted when Florida was ready, not when the calendar said to plant.

In my backyard, I planted turmeric for the first time in October. The rhizomes lay on the ground. Nothing happened.

I dug them up in January. They were rotten.

The following year, I planted turmeric in March. The rhizomes sprouted within two weeks.

In October, when the leaves yellowed, I dug up the rhizomes. I harvested three pounds of fresh turmeric from three planted rhizomes.

Turmeric punishes bad timing. When it produces heavily, you know you understand Florida's planting calendar.

Passionflower: The Plant That Tests Your Ecosystem Thinking

Passionflower produces edible fruit and supports Gulf Fritillary and Zebra Longwing butterflies. It grows as a vigorous vine. It handles heat, humidity, and poor soil.

But it requires you to accept that caterpillars will eat it.

Passionflower is a host plant. Butterfly larvae eat leaves. Caterpillars will strip entire sections of the vine bare.

If you spray or remove the caterpillars, the butterflies never appear.

Passionflower thrives when you let the caterpillars eat. The plant regrows faster than the caterpillars can strip it. The caterpillar turns into a butterfly.

The ecosystem works.

When passionflower thrives and butterflies are constant in your yard, it means you stopped trying to control pests. It means you understood some plants exist to be eaten, and that's not a failure.

That's the function.

In my backyard, I sprayed passionflower the first summer. The caterpillars died. The plant looked clean.

No butterflies appeared.

The following year, I left the caterpillars alone. They ate the plants. It looked terrible for two weeks. Then it regrew.

Butterflies appeared everywhere. The plant produced fruit.

Passionflowers don't tolerate control. When butterflies are constant and the plant produces fruit, you know you've learned to work with ecosystems instead of managing individual plants.

Moringa: The Plant That Tests Everything

Moringa is the fastest-growing edible perennial in Florida. It handles heat, humidity, and poor soil. It produces edible leaves, pods, flowers, and seeds.

It grows year-round in South Florida and regrows from the roots after freezes in Central Florida.

But moringa only thrives when you get everything right.

Plant moringa in poor drainage, and root rot kills it. Overwater it during summer, and fungus moves in. Over-fertilize it, and aphids will swarm. Plant it in spring and it struggles through the summer. Prune it wrong and it grows tall and leggy.

Moringa requires correct timing, drainage, spacing, pruning, and fertilizing. It's not a difficult plant.

But it punishes incomplete knowledge.

When moringa thrives in your yard, producing heavily and staying healthy year after year, it means you understand Florida gardening at a systems level.

It means you're no longer following individual rules. You're thinking about how timing, drainage, soil, pests, and pruning work together.

In my backyard, moringa was the plant that taught me everything. I killed it with poor drainage. I killed it with bad timing. I stressed it with overfertilization and bad pruning.

Then I planted moringa in October in a mounded area with excellent drainage. I pruned it hard every few months. I fertilized lightly in October and February. I tolerated aphids and let ladybugs handle them. I stopped watering during the summer rains.

That moringa is still alive today. It produces leaves constantly. I harvest from it every week.

It thrives because I finally learned how all the pieces fit together.

Moringa doesn't forgive incomplete knowledge. When it thrives year after year, you know you've figured Florida out.

A Quick Mastery Check

Answer these questions honestly:

Does cassava produce clean, firm tubers?

Does rosemary thrive with minimal watering?

Does turmeric produce large rhizomes at harvest?

Do butterflies dominate your passionflower, not sprays?

Does moringa stay productive year after year?

If yes to most of these, you've figured Florida out.

These plants are badges of honor

When these five plants thrive in your yard, you're no longer a beginner.

Cassava means you understand drainage. Rosemary means you learned restraint. Turmeric means you understand timing. Passionflower means you work with ecosystems. Moringa means you understand systems-level thinking.

You don't need to grow all five. But when one or more of them thrives, you have proof that you understand Florida gardening beyond following instructions.

In my backyard, these plants are the ones I'm proudest of. Not because they're rare or difficult.

Because they show me I'm no longer fighting Florida. I'm working with it.

My wife walked through the garden last fall and stopped at the moringa. She said, "Remember when you couldn't keep this alive?"

I remembered. Killed three before the fourth one stuck.

She looked at it—ten feet tall, loaded with leaves, producing constantly—and said, "You figured it out."

She was right.

But growing mastery of plants isn't the goal

The goal isn't to grow these five plants just to prove you can.

The goal is to build a perennial-first yard that produces food year after year with less work every season. A system that improves instead of restarting. A garden that requires observation, not constant intervention.

That's what the next chapter is about. Designing a perennial-first yard. How to layer sun, shade, edibles, and ornamentals into a system that works together. Developing a long-term perspective. How to build something that gets better every year instead of starting over every spring.

Skip it and you'll have individual plants that work. Read it and you'll have a food forest that produces more every year while requiring less from you.

Chapter 16 Designing a Perennial First Yard

HOW TO THINK LONG TERM. LAYERING SUN, SHADE, EDIBLES, AND ORNAMENTALS INTO A SYSTEM THAT IMPROVES EVERY YEAR INSTEAD OF RESTARTING EVERY SEASON.

M ost people design their yards one plant at a time. I planted Moringa in my backyard because of the abundance of sun. I placed the katuk in the shade and scatter other plants where space was available.

That's not design. That's filling holes.

A perennial-first yard isn't a collection of plants. It's a system.

Plants support each other. Tall plants create shade for shorter plants. Ground covers suppress weeds. Deep-rooted plants bring nutrients up while shallow-rooted plants use what's near the surface. Nitrogen-fixers feed the soil.

This is how food forests work. Layers. Functions. Relationships.

In my backyard, I spent two years planting randomly. The plants grew, but they didn't work together. I was managing individual plants, not building a system.

Then I started thinking in layers. Tall plants in the back create shade. Mid-height plants filling the space. Groundcover suppressing weeds.

The yard required less work. The plants supported each other instead of competing.

This chapter is about designing a perennial-first yard. Not a garden. A system.

One that improves every year instead of restarting every season.

But first, let me show you what my actual yard looks like. Not the theory version. The real one. The mistakes are still visible, and I had to redo some parts twice.

The Difference Between a Garden and a System

A garden requires constant input.

You plant, weed, water, fertilize, and harvest.

Then you restart. Every season. Every year. The work never decreases.

A system requires high input in year one, moderate input in year two, and minimal input in year three and beyond.

The work decreases over time because the system becomes self-maintaining.

Year One: Plants. Mulch heavily. Water regularly. Weed frequently. You're building the system.

Year Two: The plants have now established themselves. Mulch suppresses most weeds. Watering and fertilizing decrease. You're maintaining, not building.

Year Three and Beyond: Plants shade out weeds. They hold moisture. They feed the soil with leaf drop. You are constantly harvesting. You prune occasionally. Fertilize with a light touch. The system maintains itself.

In my backyard, year one was exhausting. Year two was easier. Year three, the system took over and produced more food with less work.

That's the difference. Gardens require constant work. Systems work for you.

How My Actual Yard developed (The Mistakes Nobody Talks About)

In year one, I had a flat backyard with St. Augustine grass and one massive oak tree.

I wanted a food forest. So I picked a corner, dug some holes, planted moringa and cassava in flat ground, and waited for the magic to happen.

Three months later, after a week of heavy rain, I walked outside and found standing water around every plant I'd put in the ground.

The moringa looked miserable. Yellow leaves. Stunted growth.

Two cassava plants rotted at the base and fell over.

Cost of dead plants: forty-five dollars.

The cost of learning Florida doesn't have drainage: priceless.

That's when I figured out the mound strategy.

The Mound Strategy (Or: How to Stop Drowning Your Plants)

Florida soil doesn't drain.

Especially in central Florida, where the water table sits eighteen inches below the surface during the rainy season.

Plant anything in flat ground and it sits in water for days after heavy rain.

Most tropical perennials hate wet feet. Moringa, cassava, papaya, citrus. They all need drainage, or they rot.

So you build mounds.

Here's what I did:

I picked the areas where I wanted to plant trees and large perennials. Then I built mounds twelve to eighteen inches tall and three to four feet wide.

Materials: Native soil, compost, wood chips, leaf mulch. Whatever I had available. I wasn't fancy about it. I piled material, shaped it into a gentle mound, and planted on top.

Cost: Free if you use yard waste and wood chips from tree services. Fifty to one hundred dollars if you buy compost and mulch.

The mounds drain. Water runs off the sides instead of pooling around the plant roots.

I replanted moringa on mounds. Cassava on mounds. Papaya on mounds.

All of them grew faster. No more yellow leaves. No more rot.

Now everything I plant goes on a mound unless it specifically likes wet feet. Taro, cranberry hibiscus, and ginger in the rainy season.

The part nobody tells you: mounds settle.

After six months, my eighteen-inch mounds were twelve-inch mounds. After a year, they were eight inches tall .

So I add material every year. More compost, wood chips, and leafs from the plants themselves.

The mounds stay tall enough to drain. The plants stay happy.

If you only take one thing from this chapter, take this: plant on mounds in Florida. Your success rate will double.

Water Access (Or: Why I Planted Everything Too Far from the Hose)

In year one, I designed my layout based on sun exposure and aesthetics.

Moringa in the back corner because it's tall, and I wanted it out of sight. Katuk along the fence because I thought it would look nice. Cassava in the far side yard because that's where I had space.

Then I had to water them.

The hose reaches thirty feet from the spigot. Everything I planted was forty to sixty feet away.

I hauled a watering can back and forth for three months. Four kids running around. Dinner to make. A job that doesn't care about my food forest.

I got tired. I skipped waterings. Plants stressed. Growth slowed.

Cost of that lesson: time I'll never get back and about sixty dollars in plants that struggled more than they should have.

Now I design around water access first, and sun exposure second.

Here's my current system:

Zone 1 (within thirty feet of the hose): High-maintenance plants that need regular watering. Moringa. Papaya. Katuk. Greens. Anything I'm establishing.

Zone 2 (thirty to fifty feet from the hose): Established plants that can handle irregular watering. Cassava. Lemongrass. Cuban oregano. Sweet potato vines.

Zone 3 (fifty-plus feet from the hose): Drought-tolerant plants only. Prickly pear. Agave. Plants I water maybe once a month in the dry season.

I also run a soaker hose to the back section now.

Cost: thirty-five dollars.

Time saved per week: two hours.

Best thirty-five dollars I've spent.

If you're planting a new food forest, map your hose reach before you plant anything. Put the high-value, high-harvest plants close to water. Put the tough survivors farther out.

Don't design a beautiful layout you can't actually maintain.

The Seven Layers of a Perennial Food Forest

We design food forests in layers. Each layer occupies a different vertical space. Each layer has a function.

Together, they create a dense, productive system that mimics natural forests.

You don't need to install all seven layers at once. Most successful perennial yards start with three layers and add the rest.

Layer 1: Canopy Trees (15-plus feet tall)

Large fruit and nut trees. Mango, avocado, pecan, persimmon. These create shade, moderate temperatures, and produce large harvests.

In Florida, canopy trees need space. Plant them with twenty to thirty feet between trees. Use the space beneath them for shade-tolerant perennials.

Layer 2: Understory Trees (8 to 15 feet tall)

Smaller fruit trees and large perennials. Moringa, papaya, banana, citrus on dwarfing rootstock. These fill the space between canopy trees and produce heavily.

In my backyard, moringa functions as an understory tree. It grows ten to fifteen feet tall and creates light shade. I plant katuk and longevity spinach beneath it.

The moringa shades the greens. The greens suppress weeds. Both produce food.

Layer 3: Shrub Layer (3 to 8 feet tall)

Perennial shrubs and large herbs. Katuk, chaya, elderberry, rosemary, Cuban oregano. These fill mid-height spaces and produce leaves, berries, or herbs.

Shrubs create the bulk of the productive space in most perennial-first yards. They grow fast, produce heavily and stay at chest height for easy harvesting.

Layer 4: Herbaceous Layer (6 inches to 3 feet tall)

Perennial greens and herbs. Longevity spinach, Okinawa spinach, society garlic, chives, Tulsi basil. These fill low spaces and produce constantly.

Herbaceous plants are what you harvest from daily. They grow fast, regrow after cutting, and provide the salad greens, cooking greens, and fresh herbs you use every day.

Layer 5: Groundcover Layer (0 to 6 inches tall)

Low-growing plants that spread across the ground. Sweet potato leaves, beach morning glory, perennial peanut. These suppress weeds, hold moisture, and sometimes produce food.

Layer 6: Climbing Layer (vines)

Vining plants that grow vertically. Passionflower, chayote, climbing spinach. These use vertical space without taking up ground space.

Layer 7: Root Layer (underground)

Plants grown for roots and tubers. Cassava, turmeric, ginger, sweet potatoes. These produce food underground while leaving above-ground space for other plants.

What My Seven Layers Actually Look Like (Not Theory)

Behind my house, I have a ten-by-fifteen-foot section that contains all seven layers. Took me three years to build.

Canopy: The oak tree that was already there. I didn't plant it. I'm using it.

Understory: Three moringa trees forming a loose triangle. I coppice them every year to keep them at ten feet instead of letting them hit twenty.

Shrub layer: Katuk planted in partial shade under the moringa. Cuban oregano sprawling at the edges.

Herbaceous layer: Longevity spinach and Okinawa spinach filling the mid-level space. Society garlic in clumps between larger plants.

Groundcover: Sweet potato vines covering every bare spot. I didn't plan for them to spread that much. They just did.

Climbing layer: Passionflower vine climbing the fence at the back edge. Produces fruit, I mostly give away because we get more than we can eat.

Root layer: Turmeric and ginger in the shadiest spots. Cassava along the sunny edge (planted on a mound).

This one section produces moringa leaves, katuk, multiple greens, herbs, sweet potatoes, turmeric, ginger, cassava, and passionfruit.

Same footprint as a small garden shed.

It took three years to get here. Started with just moringa and katuk. Added layers as I learned what worked.

Now it's dense. Almost too dense. I have to thin the sweet potato vines twice a year because they're trying to take over.

That's what a mature system looks like. Productive chaos.

How to Design for Sun and Shade

Most yards have areas with full sun and areas with partial shade. Instead of fighting this, design around it.

Full-sun areas: Plant heat-loving perennials. Moringa, cassava, prickly pear, agave, lemongrass, Cuban oregano, rosemary.

Partial shade areas: Plant shade-tolerant perennials. Katuk, chaya, longevity spinach, Okinawa spinach, cranberry hibiscus, sweet potato leaves.

Creating shade where you need it: If you have too much sun, plant tall perennials like moringa or bananas to create shade. Within a year, you can convert a full-sun area into partial shade.

In my backyard, I had one section that was full sun all day. Too hot for most perennials in summer.

I planted three moringa trees along the south side. Within a year, the moringa created light shade. I planted katuk and longevity spinach beneath them.

Now that section produces year-round.

But here's what I screwed up: I planted the moringa trees too close together. Six feet apart.

Within two years, their canopies overlapped. The middle tree got shaded out by the other two. It grew spindly and weak.

I had to remove it.

Cost: ten dollars for the original tree, plus the year of growth I lost.

Now I space understory trees ten to twelve feet apart minimum. They still create shade for the plants below, but they don't compete.

The Oak Tree Problem (And How I Stopped Fighting It)

That massive oak tree in my backyard drops leaves constantly. Tons of them.

For years, I raked them up and hauled them to the curb.

Then I tired of raking.

So I stopped.

I let the leaves fall where they landed. They piled up under the oak tree six inches deep.

Within three months, the bottom layers were decomposing into leaf mold. Within six months, I had rich, dark compost under that tree.

I planted shade-tolerant perennials directly into the leaf mulch. Katuk. Okinawa spinach. Turmeric. Ginger.

No soil prep. Just pushed the leaves aside, stuck the plant in the ground, pulled the leaves back around it.

Those plants grew better than anything I'd planted with careful soil amendments.

The oak tree was feeding them. The leaf drop was mulch and fertilizer in one.

Now I call that section "the Oak Guild." It's one of the most productive areas in my yard, and I do almost nothing to maintain it.

Leaves fall from the trees. The leaves feed the soil. The plants thrive.

I stopped fighting the oak tree and started using it.

If you have an enormous tree in your yard, don't curse it. Plant under it.

Spacing for Systems, Not Individual Plants

When you design a system, spacing is about function, not just plant size.

Tight spacing (12 to 18 inches): Ground covers and herbaceous plants. These fill space fast and suppress weeds.

Medium spacing (3 to 5 feet): Shrubs and mid-height perennials. These need airflow but should fill in to create a dense layer.

Wide spacing (8 to 15 feet): Large perennials and small trees. These create structure and shade for smaller plants beneath them.

The goal is to layer plants, occupying every vertical space. The tall plants in the back. Mid-height plants in the middle. Ground cover at the base.

No bare soil. No wasted space.

Here's where I messed up early on: I followed the spacing guidelines on the plant tags.

Katuk tag said, "Space 4 to 6 feet apart." So I planted three katuk plants six feet apart.

They took two years to fill in. Weeds grew in all the empty spaces between them.

Now I plant katuk three feet apart. They will fill in within six months. No room for weeds.

Although the plants crowd each other slightly, they stay healthier because they provide shade for each other and keep the soil cool.

Tight spacing works in Florida if you're layering. The plants support each other instead of competing.

Start Small and Expand

The biggest mistake people make when designing a perennial-first yard is trying to convert the entire yard at once.

Start with one area. One hundred square feet. Plant it densely. Mulch heavily. Get that area working.

Then expand.

Year One: Plant one hundred square feet. Focus all your energy on that one area.

Year Two: Expand by another one hundred square feet. We have established the first area, and it requires less work.

Year Three: Expand again. The first area maintains itself. They established the second area. You add a third.

This approach builds confidence and systems knowledge. You learn what works in your specific yard.

You make mistakes in a small area, not across the entire property.

In my backyard, I started with a ten-by-ten-foot area. I planted moringa, katuk, longevity spinach, and Cuban oregano.

The following year, I expanded by another ten-by-ten area.

Now, five years later, I have five hundred square feet of perennial production. I built it piece by piece.

The year I tried to plant two hundred square feet at once, I got overwhelmed. Half the plants died from neglect because I couldn't keep up with watering and weeding.

Cost of that failure: one hundred twenty dollars in dead plants.

Start smaller than you think you should. Get it right. Then expand.

Use edges and borders strategically

Edges are transition zones between two different environments. Lawn to garden. Sun to shade. Wet to dry.

Edges are the most productive areas in any landscape because they receive resources from both sides.

Fence line edges: Plant climbers like passionflower and chayote. Plant shrubs like katuk and Cuban oregano. These edges get reflected heat from fences and often have good drainage.

Driveway and walkway edges: Plant low-maintenance perennials that handle neglect. Society garlic, chives, longevity spinach.

Lawn to garden edges: Plant dense ground covers to create a clear border. Sweet potato leaves, perennial peanut, beach morning glory. These prevent grass from invading the garden.

Edges are where you get the most production from the least space. Use them intentionally.

In my yard, the fence line along the back gets full afternoon sun. Brutal in summer.

I planted lemongrass along that edge thinking it would handle the heat (it does) and create a privacy screen (it did).

What I didn't expect: the reflected heat from the fence made the lemongrass grow even faster. That strip produces more lemongrass than I can use.

I harvest it constantly for tea and cooking. I give stalks away to neighbors.

The fence line I thought was a problem turned into one of my highest-producing areas.

The Front Yard Reality (Productive Chaos Meets Suburban Aesthetics)

My front yard is the compromise zone.

I want food production. My neighbors want the subdivision to look normal. The HOA wants anything that isn't a lawn to be "landscaping."

So I planted productive perennials that look ornamental enough to avoid complaints.

Along the front walkway: Society garlic (purple flowers, looks decorative). Lemongrass (ornamental grass appearance). Sweet potato vines with dark purple foliage (look like decorative groundcover).

In the small bed near the mailbox: Rosemary (looks like a landscaping shrub). Cuban oregano (looks like a flowering succulent if you squint).

Behind the hedge of society garlic, Katuk. Longevity spinach. Okinawa spinach. Stuff the neighbors can't see from the street.

This setup has survived five years without HOA complaints. The neighbors think I have nice landscaping. I harvest from it constantly.

The front yard produces about thirty percent of what the backyard produces, but it's in a tenth of the space and requires almost zero maintenance.

I water it when I remember to . It keeps growing anyway.

Plan for Harvesting, Not Just Growing

Most people design for planting but forget to design for harvesting.

Plants that produce food need to be accessible. If you plant katuk in the back corner behind three other shrubs, you won't harvest from it.

Design rule: Put high-yield plants in accessible locations. Paths. Edges. Front of the planting bed. Plants you use daily should be easy to reach.

In my backyard, I used to plant randomly. Katuk in the back. Longevity spinach in narrow spaces between shrubs.

The plants produced, but I didn't harvest from them because they were hard to reach.

Now I plant high-harvest plants at the front of the beds. Katuk. Longevity spinach. Cuban oregano. Moringa within arm's reach.

I walk past them every day. I harvest constantly.

If you can't reach it easily, you won't harvest it. Design for access, not just for growth.

I learned this the hard way when I planted elderberry in the back corner behind the moringa trees. It produced berries. I saw them from the window.

But getting to them meant pushing through moringa branches and dodging katuk stems.

I harvested from that elderberry exactly twice in two years. The birds got the rest.

Eventually, I moved it to an accessible spot along the fence line. Now I actually harvest the berries.

Cost of replanting an established elderberry: one afternoon of digging, some choice words when I hit a root, and a sore back the next day.

Put your high-value plants where you can reach them without an expedition.

Paths Matter More Than You Think

In year one, I didn't build paths. I just planted densely and figured I'd walk between the plants.

That worked for about six months. Then the plants filled in, and I couldn't walk between them without trampling something.

I lost production because I couldn't access half my plants without stepping on ground covers or breaking branches.

Now I build paths first, then plant around them.

Materials: Wood chips. Mulch. Whatever's free from tree services. I laid down cardboard to suppress grass, piled wood chips on top six inches thick, and packed it down.

Paths are two feet wide minimum. Wide enough to walk comfortably with a harvest basket. Wide enough to kneel and work of crushing plants.

I have three main paths in my backyard now:

One running from the back door to the oak tree area (my most-accessed section).

One along the fence line (for accessing the cassava and lemongrass).

One cutting through the middle (connecting everything).

Total cost of paths: free wood chips from tree trimming services.

Time saved per week: at least an hour not trying to navigate through dense plantings.

Build the paths. You'll thank yourself later.

What a Mature Perennial-First Yard Looks Like

A mature perennial-first yard is dense. Plants fill every layer.

Tall plants in the back. Mid-height plants in the middle. Ground cover at the base. Vines on fences. Roots underground.

There's no bare soil. Weeds don't grow because there's no space for them.

You walk through the yard and harvest constantly. Moringa leaves from the top. Katuk from the shrub layer. Longevity spinach from the ground. Cuban oregano from the edges. Cassava from below ground.

The yard produces more food every year. Not because you're working harder. Because the system is maturing.

Roots are deeper. Plants are larger. Production increases while your work decreases.

In my backyard, this took three years. In the first year, I planted and managed constantly. After the second year, the plants established themselves, and I managed them moderately. Year three, the system took over, and I harvested constantly while managing minimally.

My wife noticed before I did. She was pulling weeds in year one almost every week. By year three, she told me she couldn't find any weeds to pull.

The plants had taken over.

But here's what maturity also means: it's messy.

Leaves everywhere. Stems crossing paths. Sweet potato vines trying to climb the fence. Passionflower seeds sprouting in random spots.

My neighbor asked me once if I was "planning to clean up the garden."

I told him it was already clean. This is what productive looks like.

He didn't get it. But my grocery bill understands perfectly.

The Layout Mistakes I'd Avoid If I Started Over

If I could redesign my yard from scratch, here's what I'd do differently:

One: Build all the mounds before planting anything. I built mounds as I went, which meant replanting things that should've been on mounds from the start.

Two: Install the soaker hose system first. Would've saved me three months of hauling water.

Three: Space understory trees wider. Ten to twelve feet instead of six. Crowded trees waste space and money.

Four: Create paths immediately. Don't wait for the plants to fill in and then realize you can't reach anything.

Five: Put the highest-producing plants closest to the back door. I planted moringa in the back corner for aesthetics. Should've planted it ten feet from the door for convenience.

Six: Start with half the number of plants and twice the mulch. Plants grow. Bare soil grows weeds. Heavy mulch first, then plants second.

Seven: Plant more nitrogen-fixers earlier. I added pigeon peas in year three. Should've planted them in year one. The soil would've been better from the start.

These mistakes cost me time and money. Maybe three hundred dollars in replanted or dead plants. Probably fifty hours of unnecessary labor.

But they taught me how to design a system that works.

Your layout will be different (and that's fine)

My yard is a third of an acre in west-central Florida with one big oak tree, a fence on three sides, and a slope toward the back.

Your yard is different. Different trees. Different sun patterns. Different soil. Neighbors who are different.

The principles remain the same:

Build mounds for drainage.

Design around water access.

Layer plants vertically.

Start small and expand.

Build paths before you need them.

Put high-yield plants where you can reach them.

But the actual layout? That's yours to figure out.

Walk around your yard at different times of the day to see where the sun hits. See where water pools. See where you actually walk.

Design around what's already there instead of fighting it.

We should use the oak tree for shade plantings. Use the fence for trellises. Use the slope for drainage.

Then plant. One section at a time.

But Design Is Only the Beginning

It is within your knowledge to layer plants in seven layers. You possess the knowledge to design for both sun and shade. You understand how to space systems. You know how to start small and expand. You know how to use edges and plan for harvesting.

You know about mounds and water access and paths and all the mistakes I made so you don't have to make them.

But design only matters if plants stay alive.

And in Florida, plants die. Not from neglect. Excessive hurricane wind or rainfall, from pests. From diseases. From random catastrophic events that no amount of planning can prevent.

The next section is about what to do when something still dies. Because it will. And when it does, you need to know how to respond without losing confidence or giving up on the system.

Closing: What to Do When Something Still Dies

BECAUSE PLANTS DIE IN FLORIDA NOT FROM NEGLECT, FROM REALITY.

You did everything right, and the plant died anyway.

This happens. Not because you failed. Because Florida is unpredictable.

You planted them in October and tested the drainage, mulched, watered, fertilized, everything.

And the plant still died.

It's possible that an unexpected freeze in December killed the moringa you had planted six weeks before. Maybe a three-week drought stressed a newly planted katuk before it established itself. Maybe armadillos dug up your cassava. Perhaps a fungal outbreak advanced more rapidly than your response allowed.

This isn't a failure. This is Florida.

The difference between someone who succeeds at Florida gardening and someone who quits isn't whether they lose plants. Everyone loses plants.

The difference is how they respond.

Quick Triage: What Killed It?

When a plant dies, figure out why before you replant.

Weather

Unexpected freezes. Flooding. Extended droughts. Heat waves beyond normal.

What to do: Wait. Don't replant immediately. Prune dead wood and see if the plant regrows from the roots. If it doesn't, replant in the right planting window (October through February). Weather deaths aren't preventable. They're just Florida.

Pests and Disease

Fungal diseases despite proper spacing. Pest outbreaks that moved faster than predators could respond. Bacterial infections with no visible symptoms.

What to do: Remove infected material. Improve airflow if necessary. Don't spray unless the problem spreads to other plants. Replant in the next planting window. Most pest and disease deaths represent isolated incidents, not systemic problems.

Water and Drainage

Root rot from standing water. Drought stress from inconsistent watering during establishment.

What to do: Stop watering. Check drainage again. If drainage is poor, mound the area or choose plants that tolerate wet soil. If drainage is good but you over-watered, adjust your watering schedule and replant.

Chemical Drift or Runoff

Lawn treatments. Weed killer drift. Fertilizer runoff from neighboring properties.

What to do: Check with neighbors about recent applications. Create buffer zones between treated areas and your garden. Replant away from potential drift zones.

Common Causes That Aren't Your Fault

Some plant deaths have nothing to do with what you did.

Weather deaths: I lost three established moringa to an unexpected January freeze. 28°F for six hours. I couldn't have prevented it.

I replanted in March. Those moringa are still alive today.

Pest and disease deaths: Fungal outbreaks that move faster than you can respond. Pest explosions before predators arrive. Bacterial infections that show no symptoms until the plant collapses.

These happen. You can't prevent all of them.

Wildlife deaths: Armadillos dig up roots. Deer eat fresh growth. Squirrels strip bark. Raccoons knock over containers. And don't even get me started on iguanas.

You can fence some areas. But mostly, you replant and move on.

How to Investigate a Death

Don't just replant in the same spot without figuring out what went wrong.

Check the roots: dig up the dead plant. Look at the roots.

Healthy roots are firm and white or tan. Rotten roots are mushy and dark.

If the roots were rotten, drainage was the problem.

Check the stem at the soil **line:** Feel the stem where it meets the soil.

If it's soft or discolored, crown rot killed it. This means water sat around the base for too long.

Check for pests: Look for aphids on fresh growth. Check for root-knot nematodes (swollen, knotted roots). Look for armadillo holes near the plants.

Check for chemical drift: Did neighbors spray recently? Did you apply lawn treatment nearby? Chemical drift kills plants fast and looks like disease.

Check the planting spot: Do the drainage test again. Conditions change. Tree roots grow into areas that used to drain well. Soil compacts. What worked two years ago might not work now.

In my backyard, I lost a katuk that had been thriving for over a year. I dug it up. The roots were completely rotten.

I did the drainage test. Water sat for four hours.

A nearby tree had grown roots into that area and blocked the drainage. The spot that worked when I planted the katuk didn't work anymore.

I moved to a different spot with better drainage. The new katuk thrived.

When to replant the same spot

Replant if:

The plant died from weather (freeze, storm, drought) that wasn't caused by the spot itself

The plant died from pests or disease that aren't specific to that location

Drainage is still good when you test it again

Sun exposure is still correct

Don't replant if:

The roots were rotten, and drainage is still poor

The spot floods regularly

The spot has become too shady or too sunny for that plant

You've lost multiple plants in the same spot

When to Try a Different Plant

Some spots won't work for certain plants. That doesn't mean the spot is bad. It means the plant and the spot don't match.

In my backyard, I tried rosemary in a low-lying area twice. Both plants died from occasional flooding.

I didn't try rosemary a third time. I planted elderberries instead.

Elderberry loves wet soil. It thrived.

The spot wasn't wrong. The plant was wrong for that spot.

How to Keep Confidence When Plants Die

I've lost dozens of plants over ten years. Some taught me lessons—poor drainage, bad timing, overfertilization.

Others taught me nothing. They died from freezes, storms, armadillos.

I kept replanting. I kept adjusting. My success rate went up.

In the first year, I lost about half of what I had planted. In year three, I lost maybe 10%. Year five, I lost almost nothing except weather casualties.

The difference wasn't that I got better plants. The difference was that I understood Florida's rules and stopped fighting them.

But I still lose plants occasionally. Last winter, an unexpected cold snap killed a papaya I'd been growing for two years.

I felt annoyed for about a day. Then I replanted in the right window. The new papaya is producing now.

Plant deaths don't erase what you know. They don't mean the system failed.

They mean you're gardening in a state where weather, pests, water, and wildlife occasionally win a round.

Your Job Now

You planted correctly. Your timing was perfect. You managed correctly. You learned the rules and stopped fighting the climate.

If a plant dies now, it doesn't mean you failed.

It means Florida happened.

Investigate. Learn. Adjust. Replant.

Some plants will die. The system will thrive.

That's what success looks like here.

Welcome to Florida gardening!

ACKNOWLEDGMENTS

T his book would not exist without the people who shaped my life, my values, and my love for growing things.

First and always, thank you to my wife, Toni. My best friend, my life partner, and my gardening partner. You have forgotten more about gardening than I will ever know, and you share that knowledge with patience, generosity, and love. Everything I build is stronger because you are beside me. And thank you to my wonderful children, Jade, Jermaine Jr., Jessi, and J'wa. I have always wanted you to be as proud of your old man as I am of the incredible kids I get to call mine.

Thank you to my mother, who introduced me to gardening at a young age and showed me, by example, what strength truly looks like. Your influence runs through every page of this book. Thank you for introducing me to things as a child that I did not fully understand but which I have deeply come to appreciate as an adult. I see your intentions much more clearly now.

Thank you to my second set of parents, Mr. and Mrs. Paige. You were there for me as a kid, and you are still there for me and my family now.

And finally, thank you to my dad, who is watching over us. Since your passing, your lessons and words have become clearer and more meaningful in my life than ever before. I miss you, Dad.

www.ingramcontent.com/pod-product-compliance
Lightning Source LLC
Chambersburg PA
CBHW060418130626
46555CB00005B/2114